"With infectious enthusiasm for his subject, Hazen introduces readers to Earth's defining moments . . . [and] argues that understanding the interplay between Earth's geological and biological pasts can help us predict and prepare for the future of life on our planet."

—Saron Yitbarek, *Discover*

"A fascinating new theory on the Earth's origins written in a sparkling style with many personal touches . . . Hazen offers startling evidence that 'Earth's living and nonliving spheres' have co-evolved over the past four billion years." —*Kirkus Reviews* (starred review)

"Concise and colorful . . . Drawing on the latest research and influenced by advances in astrobiology, Hazen takes a radical standpoint . . . to tell the amazing tale of our planet's intertwined living and nonliving spheres." —Birger Schmitz, *Nature*

"Lively and vivid . . . Hazen is a master storyteller with a great story to tell. . . . A sweeping rip-roaring yarn of immense scope, from the birth of the elements in stars to meditations on the future habitability of our world. . . . Anyone new to earth history will find Hazen's account a revelation." —A. D. Anbar, *Science*

"I'm not competent to assess the accuracy of Robert Hazen's thesis about geological and biological history, but I am competent to judge it a fascinating story, far more alive than you might guess if all you knew was the subject was old dead (?!) rocks."
—Bill McKibben, author of *Eaarth: Making a Life on a Tough New Planet*

"Hazen takes us on one of the grandest tours of them all—the 4.5 billion year history of our planet. From the atoms of the crust of the Earth come our bodies, the entire living world, and this exciting book. Read Hazen and you will not see Earth and life in the same way again."
—Neil Shubin, paleontologist and author of *Your Inner Fish*

PENGUIN BOOKS

THE STORY OF EARTH

Robert M. Hazen is the Clarence Robinson Professor of Earth Science at George Mason University and a senior scientist at the Carnegie Institution's Geophysical Laboratory. The author of numerous books—including the bestselling *Science Matters*—he lives with his wife in Glen Echo, Maryland.

THE STORY OF
EARTH

The First 4.5 Billion Years,
from Stardust to Living Planet

ROBERT M. HAZEN

PENGUIN BOOKS

PENGUIN BOOKS
Published by the Penguin Group
Penguin Group (USA) Inc., 375 Hudson Street,
New York, New York 10014, USA

USA | Canada | UK | Ireland | Australia | New Zealand | India | South Africa | China
Penguin Books Ltd, Registered Offices: 80 Strand, London WC2R 0RL, England
For more information about the Penguin Group visit penguin.com

First published in the United States of America by Viking Penguin,
a member of Penguin Group (USA) Inc., 2012
Published in Penguin Books 2013

THE LIBRARY OF CONGRESS HAS CATALOGED THE HARDCOVER EDITION AS FOLLOWS:
Hazen, Robert M., 1948–
    The story of Earth : the first 4.5 billion years, from stardust to living planet / Robert M. Hazen.
        p. cm.
    Includes index.
    ISBN 978-0-670-02355-4 (hc.)
    ISBN 978-0-14-312364-4 (pbk.)
    1. Earth.   I. Title.
    QE501.H325 2012
    550—dc23   2011043713

Printed in the United States of America
20   19   18   17

Designed by Carla Bolte
Time Lines by Jeffrey L. Ward

To Gregory:

Change will come;

may you have the wisdom and courage to adapt

CONTENTS

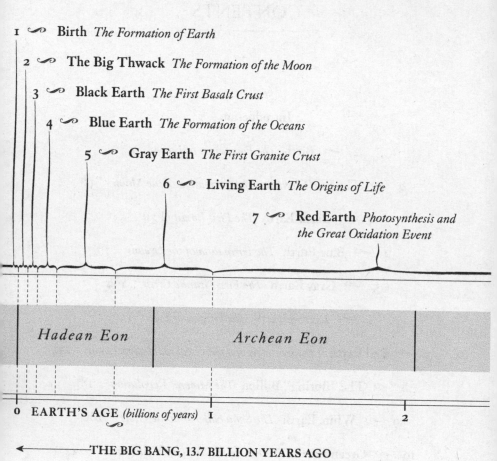

Hadean Eon   *Archean Eon*

0 EARTH'S AGE (*billions of years*)   **1**   **2**

◄── **THE BIG BANG, 13.7 BILLION YEARS AGO**

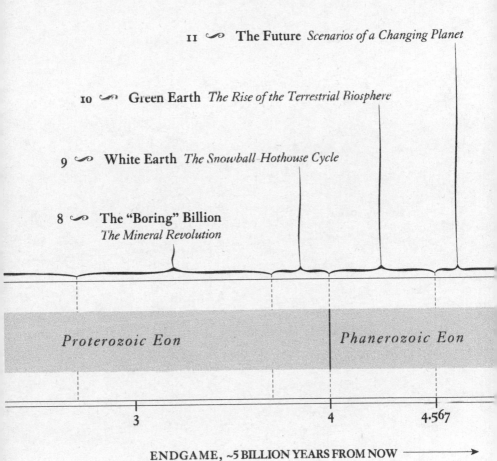

Proterozoic Eon

Phanerozoic Eon

3   4   4.567

ENDGAME, ~5 BILLION YEARS FROM NOW ⟶

THE STORY OF EARTH

Introduction

ᥫ One of the most arresting images of the twentieth century is a photo of Earthrise, taken in 1968 by a human traveler in orbit around the Moon. We have long known how precious and special our world is: Earth is the only known planet with oceans of water, with an atmosphere rich in oxygen, with life. Nevertheless, many of us were unprepared for the breathtakingly stark contrast between the utterly hostile lunar landscape, the lifeless black void of space, and our enticing marbled white-on-blue home. From that distant vantage point, Earth appears alone, small, and vulnerable but also more beautiful by far than any other object in the heavens.

We are justifiably captivated by our home world. More than two centuries before the birth of Christ, the polymath Greek philosopher Eratosthenes of Cyrene conducted the earliest documented experiment on Earth. He based his ingenious measurement of Earth's circumference on the simple observation of shadows. In the town of Syene, Egypt, at noon on the summer solstice, he observed the Sun directly overhead. A vertical post cast no shadow. By contrast, on the same day at the same time in the coastal city of Alexandria, some 490 miles to the north, a similar vertical post cast a slight shadow, thus revealing that the Sun was not quite directly overhead in that location. Eratosthenes employed the geometrical theorems of his Greek predecessor Euclid to conclude that Earth must be a sphere, and he calculated that it must have a circumference of about 25,000 miles—remarkably close to the modern established value of 24,902 miles around the Equator.

Throughout the centuries, thousands of other scholars, a few with names revered but most lost to history, have probed and pondered our planetary home. They have asked how Earth was formed, how it moves through the heavens, what it's made of, and how it works. And most of all, these men and women of science have wondered how our dynamic planet evolved, how it became a living world. Today, because of our remarkable and cumulative knowledge, and because of the wonders of human technology, we know more about Earth than ancient philosophers could have fathomed. Of course we don't know everything, but our understanding is rich and deep.

And while our knowledge of Earth has increased since the dawn of humankind, refined over the millennia into a fixed understanding, much of that progress has revealed that the study of Earth is the study of change.

Many lines of observational evidence point to the fluctuating nature of Earth year by year, epoch by epoch. Rhythmically layered or varved sediment deposits in some glacial lakes in Scandinavia display more than thirteen thousand years of alternating coarser and finer particles—the consequences of swifter erosion during annual spring thaws. Frozen glacial drill cores from Antarctica and Greenland reveal more than eight hundred thousand years of seasonal ice accumulations. And deposits of paper-thin sediment layers from Wyoming's Green River Shale preserve more than a million years of annual events. Each of those layerings rests atop vastly older rocks, which themselves hint at grand cycles of change.

Measurements of gradual geological processes point to even more immense spans of Earth history. The formation of the massive Hawaiian Islands required slow and steady volcanic activity, successive lava blankets piling up over tens of millions of years. The Appalachians and other ancient rounded mountain ranges are the result of hundreds of millions of years of gradual erosion punctuated by great landslides. The sometimes herky-jerky motions of tectonic plates have shifted continents, elevated mountains, and opened oceans over the course of geological history.

Earth has always been a restless, evolving planet. From core to crust, it is incessantly mutable. Even today the air, the oceans, and the land are changing, perhaps at a pace unequaled in our planet's recent past. We would be foolish not to care about these unsettling global changes, and indeed for many of us, it seems impossible that we wouldn't—our curiosity and care for our home comes as naturally to us as it did to Eratosthenes. But we would be equally foolish to address the current state of Earth without taking full advantage of what it already tells us about its surprising storied past, about its unpredictable dynamic present, and about ourselves and our place in its future.

Most of my life has been spent trying to understand our vibrant, complex, changeable home. As a boy, I collected rocks and minerals, cramming my room full of fossils and crystals, side by side with random bugs and bones. My entire professional career has followed that Earth-centric theme. I began with experiments at the submicroscopic scale of atoms, studying the molecular structure of rock-forming minerals, heating and squeezing tiny mineral grains to document the pressure-cooker-like effects of Earth's deep interior.

With time, my view expanded to the grander spatial and temporal tapestry of geology. From the deserts of North Africa to the ice fields of Greenland, from the shores of Hawaii to the peaks of the Rockies, from the Great Barrier Reef in Australia to ancient fossilized coral reefs in a dozen nations, Earth's natural libraries reveal a multibillion-year story of coevolution shared by elements, minerals, rocks, and life. As my research program shifted to the plausible roles of minerals in life's ancient geochemical origins, I have reveled in studies that suggest that the coevolution of life and minerals through Earth history is even more striking than previously imagined—that not only do certain rocks arise from life, as evident in limestone caverns across the continent, but that life itself may have arisen from rocks. Over four billion years of Earth history, the evolutionary stories of minerals and life—geology and biology—have intertwined in astonishing ways that are only now coming into focus. In 2008 these ideas culminated in an unconventional paper on "Mineral Evolution"—a controversial new argument that some welcomed as perhaps the first paradigm shift in mineralogy in two centuries, while others viewed it warily, as a possibly heretical reframing of our science in the context of deep time.

The ancient discipline of mineralogy, though absolutely central to everything we know about Earth and its storied past, has been curiously static and detached from the conceptual vagaries of time. For more than two hundred years, measurements of chemical composition, density, hardness, optical properties, and crystal structure have been the meat and potatoes of the mineralogist's livelihood. Visit any natural history museum, and you'll see what I mean: gorgeous crystal specimens arrayed in case after glass-fronted case, with labels showing name, chemical formula, crystal system, and locality. These most treasured fragments of Earth are rich in historic context, but you will likely search in vain for any clue as to their birth ages or subsequent geological transformations. The old way all but divorces minerals from their compelling life stories.

That traditional view has to change. The more we examine Earth's rich rock record, the more we see how the natural world, both living and nonliving, has transformed itself again and again. Our growing understanding of the twin planetary realities of time and change has made it possible to conjecture not just *how* minerals first came to be, but *when*. And the recent discoveries of organisms in places long considered inhospitable—in superheated volcanic vents, acidic pools, Arctic ice, and stratospheric dust—have enlisted mineralogy as a key discipline in the quest to comprehend the origins and survival of life. In the November 2008 issue of the field's flagship journal, *American Mineralogist,* my colleagues and I proposed a new way to think about the mineral kingdom and its incredible transformations through the unexplored dimension of time. We emphasized that many billions of years ago, there were no minerals anywhere in the cosmos. No crystalline compounds could have formed, much less survived, in the superheated maelstrom following the Big Bang. It took a half-million years for even the first atoms—hydrogen, helium, and a bit of lithium—to emerge from the cauldron of creation. Millions more years passed while gravity coaxed these primordial gaseous elements into the first nebulas, and then collapsed the nebulas into the first hot, dense, incandescent stars. Only when those first stars exploded into supernova brilliance, when expanding, cooling envelopes of element-rich gas condensed the first tiny crystals of diamond, could the cosmic mineralogical saga have begun.

And so I have become a compulsive reader of the testimony of the rocks—the compelling, if sometimes fragmentary and ambiguous, stories they have to tell of birth and death, of stasis and flux, of origins and evolution. This untold grand and intertwined tale of Earth's living and nonliving spheres—the coevolution of life and rocks—is utterly amazing. It must be shared, because we *are* Earth. Everything that gives us shelter and sustenance, all the objects we possess, indeed every atom and molecule of our flesh-bound shells, comes from Earth and will return to Earth. To know our home, then, is to know a part of ourselves.

Earth's story must be shared, too, because our oceans and atmospheres are changing at a rate rarely matched in its long history. Oceans are rising, while they are becoming warmer and more acidic. Global patterns of rainfall are changing, while the atmosphere is becoming more turbulent. Polar ice is melting, tundra is thawing, and habitats are shifting. As we will explore in the pages ahead, Earth's story is a saga of change, but on those rare previous occasions when change occurred with such alarming rapidity, life appears to have paid a terrible toll. If we are to act thoughtfully and in time for our own sake, we must become intimate with Earth and her story. For as was sublimely evident from that wondrous picture snapped from a lifeless world 239,000 miles away, we have no other home.

In the tradition of Eratosthenes and the thousands of curious minds that have followed him, my purpose in this book is to convey Earth's long history of change. As immediate and familiar as Earth might seem to be, its vibrant story embraces a succession of transformational events almost beyond imagining. To truly know your planetary home and comprehend the aeons that shaped it, you must first wrap your mind around seven core truths.

1. Earth is made of recycled and recycling atoms.
2. Earth is immensely old when compared with human time frames.
3. Earth is three-dimensional, and most of the action is hidden from view.
4. Rocks are the record keepers of Earth history.
5. Earth systems—rocks, oceans, atmosphere, and life—are complexly interconnected.
6. Earth history encompasses long periods of stasis punctuated by sometimes sudden, irreversible events.
7. Life has changed and continues to change Earth's surface.

These concepts of Earth's existence frame the intricately layered stories of atoms, minerals, rocks, and life in their vast epics of space and time; they will reappear in the following pages, in every phase of the universe's fiery origin and Earth's protracted evolution. The co-evolution of Earth and life, the new paradigm that lies at the heart of this book, is part of an irreversible sequence of evolutionary stages reaching back to the Big Bang. Each stage introduced new planetary processes and phenomena that would ultimately resculpt our planet's surface again and again, inexorably paving the way for the wondrous world we inhabit today. This is the story of Earth.

EARTH'S AGE *(billions of years)*

0	1	2	3	4	4.567
Hadean Eon	Archean Eon	Proterozoic Eon		Phanerozoic Eon

Chapter 1

Birth

The Formation of Earth

Spanning the billions of years before Earth's formation

∾ In the beginning, there was no Earth, or any Sun to warm it. Our Solar System, with its glowing central star and varied planets and moons, is a relative newcomer to the cosmos—a mere 4.567 billion years old. A lot had to happen before our world could emerge from the void.

The stage was set for our planet's birth much, much earlier, at the origin of all things—the Big Bang—about 13.7 billion years ago, by the latest estimates. That moment of creation remains the most elusive, incomprehensible, defining event in the history of the universe. It was a singularity—a transformation from nothing to something that remains beyond the purview of modern science or the logic of mathematics. If you would search for signs of a creator god in the cosmos, the Big Bang is the place to start.

In the beginning, all space and energy and matter came into existence from an unknowable void. Nothing. Then something. The concept is beyond our ability to craft metaphors. Our universe did not suddenly appear where there was only vacuum before, for before the Big Bang there was no volume and no time. Our concept of nothing implies emptiness—before the Big Bang there was nothing to be empty in.

Then in an instant, there was not just something, but everything that would ever be, all at once. Our universe assumed a volume smaller than an atom's nucleus. That ultracompressed cosmos began as pure homogeneous energy, with no particles to spoil the perfect uniformity. Swiftly the universe expanded, though not into space or anything else outside it (there is no *outside* to our universe). Volume itself, still in the form of hot energy, emerged and grew. As existence expanded, it cooled. The first subatomic particles appeared a fraction of a second after the Big Bang—electrons and quarks, the unseen essence of all the solids, liquids, and gases of our world, materialized from pure energy. Soon thereafter, still within the first fraction of the first cosmic second, quarks combined in pairs and triplets to form larger particles, including the protons and neutrons that populate every atomic nucleus. Things were still ridiculously hot and remained so for about a half-million years, until the ongoing expansion eventually cooled the cosmos to a few thousand degrees—sufficiently cold for electrons to latch on to nuclei and form the first atoms. The overwhelming majority of those first atoms were hydrogen—more than 90 percent of all atoms—with a few percent helium and a trace of lithium thrown in. That mix of elements formed the first stars.

First Light

Gravity is the great engine of cosmic clumping. A hydrogen atom is a little thing, but take one atom and multiply it by ten to the sixtieth power (that's a trillion-trillion-trillion-trillion-trillion hydrogen atoms) and they exert quite an impressive collective gravitational force on one another. Gravity pulls them inward to a common center, forming a star—a giant gas ball with epic pressures at the core. As an immense hydrogen cloud collapses, the star-forming process transforms the kinetic energy of moving atoms to the gravitational potential energy of their clustered state, which translates into heat once more—the same violent process that occurs when an asteroid impacts Earth, but with vastly more energy release. The core of the hydrogen sphere eventually reaches temperatures of millions of degrees and pressures of millions of atmospheres.

Such temperatures and pressures trigger a new phenomenon called nuclear fusion reactions. Under these extreme conditions, the nuclei of two hydrogen atoms (each with one proton) collide with such force that neutrons are transferred from one nucleus to another, making some hydrogen atoms more massive than others. After several such collisions, a helium nucleus with two protons forms. Surprisingly, the resulting helium atom is about 1 percent less massive than the original hydrogen atoms from which it formed. That lost mass converts directly to heat energy (just as it does in a hydrogen bomb), which promotes even more nuclear fusion reactions. The star "ignites," bathing its surroundings with radiant energy, while becoming ever richer in helium at the expense of hydrogen.

Large stars, many of them much bigger than our Sun, eventually used up the prodigious supplies of hydrogen in their cores. But extreme interior pressure and heat continued to promote nuclear fusion. Two-proton helium atoms in a stellar core fused to make carbon, the vital element of life with its six protons, even as new pulses of nuclear energy triggered hydrogen fusion in a spherical layer of atoms surrounding the core. Then core carbon fused to make neon, neon to make oxygen, then magnesium, silicon, sulfur, and on and on. Gradually the star developed an onionlike structure, with layer upon concentric layer of fusion reactions. Faster and faster these reactions occurred, until the ultimate iron-producing phase lasted no more than a day. By this point in the first stars' life cycles, many millions of years after the Big Bang, most of the first twenty-six elements in the periodic table had been brought into existence by nuclear fusion within many individual stars.

Iron is as far as this nuclear fusion process can go. When hydrogen fuses to produce helium, when helium fuses to produce carbon, and during all the other fusion steps, abundant nuclear energy is released. But iron has the lowest energy of any atomic nucleus. As when a blazing fire transforms every bit of fuel to ash, all the energy has been used up. Iron is the ultimate nuclear ash; no nuclear energy can be extracted by fusing iron with anything. So when the first massive star produced its inevitable iron core, the game was over, the results catastrophic. Until that point, the star had sustained a stable equilibrium, balancing its two great inner forces: gravity pulling mass toward the center, nuclear reactions pushing mass outward from the center. When the core filled with iron, however, the outward push just stopped, and gravity took over in an instant of unimaginable violence. The entire star collapsed inward with such swiftness that it rebounded off itself and exploded in the first supernova. The star was ripped apart, blasting most of its mass outward.

The Birth of Chemistry

For those readers who seek design in the cosmos, supernovas are nearly as good a place to start as the Big Bang. To be sure, the Big Bang led inevitably to hydrogen atoms, and hydrogen atoms just as inexorably produced the first stars. Yet it's not at all obvious how stars, by themselves, get you to our modern living world. A big ball of hydrogen, even if it does have a growing core-bound collection of heavier elements up to iron, doesn't seem to help move things along in a very interesting way.

But when the first big stars exploded, cosmic novelty ensued. These fractured bodies seeded space with the elements they had created.

Carbon, oxygen, nitrogen, phosphorus, and sulfur—the elements of life—were especially abundant. Magnesium, silicon, iron, aluminum, and calcium, which dominate the compositions of many common rocks and form a large fraction of the mass of Earth-like planets, also abounded. But in the incomprehensibly energetic environment of these exploding stars, these elements fused in new and exotic ways to make *all* the periodic table—elements way beyond number twenty-six. So appeared the first traces of many rarer elements: precious silver and gold, utilitarian copper and zinc, toxic arsenic and mercury, radioactive uranium and plutonium. What's more, all these elements were hurled out into space, where they could find one another and clump together in new and interesting ways through chemical reactions.

Chemistry happens when one everyday atom bumps into another. Every atom has a tiny but massive central nucleus with a positive electric charge, surrounded by a cloudlike distribution of one or more negatively charged electrons. Isolated atomic nuclei almost never interact, except in the ultimate pressure-cooker environments of stellar interiors. But electrons from one atom are constantly bumping into the electrons of adjacent atoms. Chemical reactions occur when two or more atoms meet and their electrons interact and rearrange. Such shuffling and sharing of electrons occurs because certain combinations of electrons, notably collections of two or ten or eighteen electrons, are particularly stable.

The first chemical reactions following the Big Bang produced molecules—small clusters of a few atoms tightly bound into a single unit. Even before hydrogen atoms began fusing together in stars to form helium, hydrogen molecules (H_2), each with two hydrogen atoms chemically bonded together, formed in the vacuum of deep space. Each hydrogen atom carries only one electron, which is a rather unstable situation in a universe where two electrons is a magic number. So when two hydrogen atoms meet, they pool their resources to form a molecule with that magic number of two shared electrons. Given the abundance of hydrogen following the Big Bang, hydrogen molecules surely predated the first stars and have been a perpetual feature of our cosmos since atoms first appeared.

Following the first supernova, as a variety of other elements seeded space, lots of other interesting molecules could form. Water (H_2O), with two hydrogen atoms bonded to an oxygen atom, was one early example. Chances are that nitrogen (N_2), ammonia (NH_3), methane (CH_4), carbon monoxide (CO), and carbon dioxide (CO_2) molecules also enriched the space around supernovas. All of these molecular species would come to play key roles in the formation of planets and in the origins of life.

Then came the minerals—microscopic solid volumes of chemical perfection and crystalline order. The first minerals could have formed only where the densities of mineral-forming elements were high enough, but temperatures cool enough, for atoms to arrange themselves in little crystals. Just a few million years after the Big Bang, the expanding, cooling envelopes of the first exploding stars provided the perfect settings for such reactions. Tiny crystallites of pure carbon—diamond and graphite—were probably the first minerals in the universe. Those pioneering crystals were like a fine dust, the individual grains indiscernibly small but perhaps large enough to add a bit of diamond sparkle to space. These crystalline forms of carbon were soon joined by other high-temperature solids that featured the more common elements, including magnesium, calcium, silicon, nitrogen, and oxygen. Some were familiar minerals like corundum, the chemical compound of aluminum and oxygen that is so valued in its richly colored varieties, ruby and sapphire. Tiny amounts of the magnesium silicate olivine, the semiprecious birthstone of August, also appeared, joined by moissanite, a silicon carbide often sold these days as a cheap synthetic substitute for diamond. Altogether the interplanetary dust hosted perhaps a dozen common "ur-minerals." And so, with the explosion of the first stars, the universe began to get more interesting.

Nothing happens only once in our universe (except perhaps for the Big Bang). Scattered debris of old exploded stars were constantly subjected to the organizing force of gravity. Thus remnants of the former stellar generations inexorably seeded new populations of stars by forming new nebulas, each a vast interstellar cloud of gas and dust representing the wreckage of many prior stars. Each new nebula was more iron rich, and a little poorer in hydrogen, than the one before. For 13.7 billion years, this cycle has continued, as old stars produce new stars and slowly alter the composition of the cosmos. Countless billions of stars have emerged in countless billions of galaxies.

Cosmic Clues

Once upon a time, five billion years ago, our future real estate in the galactic suburbs lay halfway out from the Milky Way's center, at the uninhabited edge of a star-studded spiral arm. Little was to be found in that unassuming neighborhood, apart from a great nebula of gas and icy dust stretching light-years across the dark void. Nine parts in ten of that cloud were hydrogen atoms; nine parts in ten of what remained were helium atoms. Ice and dust, rich in small organic molecules and microscopic mineral grains, accounted for the remaining 1 percent.

A nebular cloud in space can last many millions of years before a trigger—a shock wave from a nearby exploding star, for example—begins its collapse into a new star system. Almost 4.6 billion years ago, such a trigger initiated our Solar System. Ever so slowly, over the course of a million years, the swirling mass of presolar gas and dust was drawn inward. Like a twirling ice-skater, the big cloud rotated faster and faster as gravity pulled its wispy arms to the center. As it collapsed and spun faster, the cloud became denser and flattened into a disk with a growing central bulge—the nascent Sun. Larger and larger grew that greedy hydrogen-rich central ball, which ultimately swallowed 99.9 percent of the cloud's mass. As it grew, internal pressures and temperatures rose to the fusion point, igniting the Sun.

Clues to what happened next are preserved in the record of our Solar System—its planets and moons, its comets and asteroids, and its abundant and varied meteorites. One striking feature is that all the planets and moons orbit the Sun on the same plane, and in the same direction. What's more, the Sun and most of the planets rotate on their axes in more or less that same plane and direction. Nothing in the laws of motion requires this commonality of spin; planets and moons could orbit and rotate any which way—north to south, east to west, top to bottom, bottom to top—and still obey the law of gravity. If planets and moons were captured from distant, random sources, one might expect such a hodgepodge. The observed orbital near-uniformity in our Solar System, by contrast, suggests that planets and moons all coalesced from the same flat rotating disk of dust and gas at more or less the same time. All of these grand objects preserve the same sense of rotation—the shared angular momentum of the entire Solar System—from the time of the original swirling cloud.

A second clue to the Solar System's origins is found in the distinctive distribution of its eight major planets. The four planets closest to the Sun—Mercury, Venus, Earth, and Mars—are relatively small rocky worlds composed mostly of silicon, oxygen, magnesium, and iron. Dense rocks, like black volcanic basalt, dominate their surfaces. By contrast, the outer four planets—Jupiter, Saturn, Uranus, and Neptune—are gas giants made primarily of hydrogen and helium. These immense spheres have no solid surfaces, just an ever-thickening atmosphere the deeper you go. This dichotomy of worlds suggests that early in the history of the Solar System, within a few thousand years of the Sun's birth, an intense solar wind blew leftover hydrogen and helium far out to the colder realms. Sufficiently far from the radiant Sun, these volatile gases could cool, condense, and gather into spheres of their own. By contrast, the coarser, mineral-rich grains of dust that remained closer to the hot central star quickly clumped together to form the rocky inner planets.

Details of the violent processes that formed Earth and the other inner planets are beautifully preserved in the amazingly diverse varieties of meteorites. It's a bit unsettling to think that our home is constantly being peppered by stones falling from the sky. In fact, the scientific community didn't pay them much heed until about two hundred years ago, though there was certainly no shortage of colorful meteorite anecdotes in folklore (including several tales involving unfortunate French peasants). Even when scholars began to describe meteorite falls more formally, little in the way of reproducible scientific evidence could be mustered to document them, much less explain their provenance. The American statesman and naturalist Thomas Jefferson, upon reading the technical report from Yale University of an observed meteorite impact in Weston, Connecticut, quipped: "I find it easier to believe that two Yankee professors would lie than that stones would fall from heaven."

Two centuries and tens of thousands of meteorite finds later, their veracity is no longer in question. As meteorite experts cover more ground, and as avid collectors vie for the rarest types, museum and private collections around the world have swelled. For a time, these repositories were skewed in favor of distinctive iron meteorites, whose black crusts, weirdly sculpted shapes, and unusually high density made them stand out against everyday rocks. But the 1969 discovery of thousands of meteorites lying on pristine Antarctic ice fields changed that perception.

Meteorites are telltale clues to our planet's origins. The most common and ancient ones, the 4.566-billion-year-old chondrites, date from the time just before the planets and moons of the Solar System formed, when the Sun's nuclear reactor first turned on and intense radiant energy broiled the encircling nebula. The blast furnace effect melted the dusty disk into clots of small, sticky rock droplets, called chondrules, after the ancient Greek word for "grain." Ranging from the size of BBs to that of small peas, these products of the Sun's refining fire were melted multiple times, in repeated pulses of radiation that transformed the regions closest to the Sun. Clusters of these ancient chondrules, cemented together by finer-grained presolar dust and mineral fragments, compose the primitive chondrites that have landed on the Earth by the millions. Chondrites provide our best view of the brief time just after the Sun was born but before the planets were formed.

A second, younger class of meteorites, collectively dubbed achondrites, dates from the time when the earliest materials of the Solar System were being reworked—melted, smashed, and otherwise transformed. The diversity of achondritic meteorites is astonishing—nuggets of shiny metal, chunks of blackened rock, some as fine-grained as glass, others with lustrous crystals an inch across. Important discoveries of new varieties are still being made in some of Earth's most remote regions.

The continent of Antarctica holds vast plains of ancient blue ice—places where it never snows. Rocks that have fallen from space just lie there, dark, out-of-place objects waiting to be retrieved. International treaties banning commercial exploitation of the area, coupled with limited access to the remote ice fields, ensure that these extraterrestrial resources will be preserved for scientific study. Teams of warmly bundled scientists in helicopters and on snowmobiles systematically scour mile after square mile of these forbidding ice deserts. They carefully record and package each find, making sure that no hand, no breath contaminates its surface. Upon returning to civilization after each Antarctic summer season, these meteorite hunters deliver their treasures to public collections, most notably the Smithsonian Institution storage facilities in suburban Suitland, Maryland, where many thousands of specimens are preserved in ultraclean, airtight storage cabinets within buildings the size of football fields.

Equally rich in meteorites, though far less conducive to organized recovery and sterile curation, are Earth's great deserts in Australia, the American Southwest, the Arabian Peninsula, and most dramatically, North Africa—the vast Sahara Desert. Word has spread among Sahara-crossing nomads—Tuaregs, Berbers, Fezzanis—that meteorites can be valuable. A single precious lunar meteorite found somewhere in the shifting sands of North Africa early in the twenty-first century is reputed to have fetched a million dollars in a private sale. It's easy enough for a desert rider to get down off his camel and carry an odd stone to the next village, where someone from an unofficial guild of meteorite middlemen, networked by satellite phones and skilled in hyperbole, will offer him a pittance in cash. From one dealer to the next, bags of rocks are passed, each time with a markup, until they reach Marrakech, Rabat, or Cairo and thence travel to the buyers on eBay and the big international rock and mineral shows.

More than once on geology trips to remote parts of Morocco, I've been offered burlap bags filled with ten or twenty pounds of rocks purported to be meteorites—"no middlemen, fresh from the desert, just found last week." These cash-only "deals" are often brokered in dingy, windowless back rooms of tan mud-brick houses, away from the blazing desert Sun, where it's almost impossible to see what's being offered. Once the formalities of greeting and the traditional cups of mint tea have been shared, the seller dumps the contents on a carpet. Some of the rocks are just rocks. Ballast. It's like a test to see if you know your stuff. A few will be the commonest sort of chondrite, the size of an olive or an egg, some with a nicely melted fusion crust, the fiery result of falling fast through the sky. The starting price is always way too high. If you say they're too common, a second, smaller bag may appear, perhaps with an iron meteorite or something even more exotic.

I recall one deal worked by our guide, Abdullah, on a dusty side road a few miles east of Skoura. The seller, a distant acquaintance of questionable integrity, called by satellite and demanded secrecy. "It might be a Martian," he told Abdullah. "Nine hundred grams. Just twenty thousand dirhams." About $2,400—if it was real, if it could be added to the two dozen or so known meteorites that came from Mars, it'd be a bargain. They arranged the time and place. Two nondescript cars pulled up beside each other; three of us got out and stood in a tight circle. The rock in question was lovingly slipped from a velvet pouch. But it looked like an ordinary rock (as do all Martian meteorites). The price dropped to fifteen thousand dirhams. Then twelve thousand. But there was no way to be sure, so we passed. Later Abdullah confided to me that he had been tempted, but there are always more meteorites. It's best not to be too greedy with one big score; no one tells the truth, and all deals are final.

As in Antarctica, the equatorial deserts reveal the natural distribution of all kinds of meteorites, providing unrivaled clues to the character of the early Solar System and thus the origins of our own planet. Sadly, unlike meteorites from Antarctica, most of these specimens will never make it to museum collections for at least two reasons. First and foremost, the growing community of amateur collectors (fueled by a few wealthy aficionados and the readily available Saharan finds) is intensely competitive. Anything rare sells quickly and for a lot of money. Some of those specimens will surely wind up as donations to museums, but most are poorly handled, and much of the scientific value in a pristine find is soon lost by contamination from unprotected hands, multipurpose cloth bags, and the ubiquitous camel dung. Equally troubling is the lack of any useful documentation, as to when or where in the desert the meteorites were found. All the dealers will say is "Morocco," which is usually a falsehood, as most of the sandy Sahara lies to the east, in Algeria and Libya—countries from which it is now illegal to import specimens. So without rigorous documentation, most museums simply will not accept "Moroccan" or "North African" meteorites.

In those hostile, arid terrains of the Sahara, or the ice fields of Antarctica, any rock stands out as a foreign object fallen from the sky. Such an unadulterated sampling of the meteorite population gives scientists their best view of the earliest stages of the Solar System in which Earth formed. Chondrites represent almost nine out of every ten finds; the rest are the diverse achondrites, belonging to the few-million-years era when our young Solar System was a turbulent nebula, in which chondrites clumped together into larger and larger bodies: first the size of your fist, then the size of your car, then a small city—billions of objects a few miles or so in diameter all vying for space in the same narrow ring around the young Sun.

Larger and larger they grew: to the size of Rhode Island, then Ohio, Texas, Alaska. As thousands of such planetesimals underwent this chaotic process of accretion, they diversified in new ways. As they grew to fifty miles or more in diameter, two coequal sources of heat compounded. The gravitational potential energy of many small objects smashing together was matched in intensity by the nuclear energy of fast-decaying radioactive elements like hafnium and plutonium. The minerals making up these planetesimals were thus transformed by heat, while their interiors melted outright, differentiating into an egglike arrangement of distinctive mineral zones: a dense metal-rich core (analogous to the egg's yolk), the magnesium silicate mantle (the egg white), and the thin brittle crust (the shell). The largest planetesimals were altered by internal heating, by reactions with water, and by the intense shock of frequent collisions in the crowded solar suburbs. Perhaps three hundred different mineral species arose as a consequence of such dynamic planet-forming processes. Those three hundred minerals are the raw materials from which every rocky planet must form, and all of them are still found today in the diverse suites of meteorites that fall to Earth.

From time to time, when two big planetesimals smashed together with sufficient force, they were blasted to smithereens. (This violent process continues to this day in the Asteroid Belt beyond Mars, thanks to the gravitational disruptions from the giant planet Jupiter.) Consequently, most of the diverse achondrite meteorites we find today represent different parts of destroyed miniplanets. Analyzing achondrites is thus a bit like a messy anatomy lesson from an exploded cadaver. It takes time, patience, and a lot of bits and pieces to get a clear picture of the original body.

The dense metallic cores of planetesimals, which wound up as a distinctive class of iron meteorites, are the easiest to interpret. Though once thought to be the most common type of meteorites, the unbiased Antarctic sampling reveals that irons represent a modest 5 percent of all falls. Planetesimal cores must have been correspondingly small.

The contrasting silicate-rich mantles of planetesimals are represented in a host of exotic meteorite types: howardites, eucrites, diogenites, ureilites, acapulcoites, lodranites, and more—each of distinctive composition, texture, and mineralogy, and most named for the locality where the earliest known example was recovered. Some of these meteorites are close analogues to rock types found on Earth today. Eucrites represent a rather typical kind of basalt—the rock type that spews out of the Mid-Atlantic Ridge and blankets the ocean floor. Diogenites, composed primarily of magnesium silicate minerals, appear to be the result of crystal settling in a large underground magma reservoir. As the magma cooled, crystals more dense than their surrounding hot liquid grew and then sank to the bottom to form a concentrated mass, just as they do in magma chambers deep inside Earth today.

Occasionally, during a particularly destructive collision, a meteorite happened to snatch a piece of a planetesimal's core-mantle boundary, where chunks of silicate minerals and iron-rich metals coexisted. The result is a beautiful pallasite—a stunning mixture of shiny metal and golden crystals of olivine. Thin polished slabs of pallasite, with light reflecting off the metal and passing through the olivine like stained glass, are among the most prized specimens in the world of meteorite collecting.

As gravity clumped the early chondrites together—and as crushing pressure, scalding temperature, corrosive water, and violent impacts reworked the growing planetesimals—more and more new minerals emerged. All together, more than 250 different minerals have been found in all the varieties of meteorites—a twenty-fold increase over the dozen presolar ur-minerals. These varied solids, which include the first fine-grained clays, sheetlike mica, and semiprecious zircon, became the building blocks of Earth and other planets. Bigger and bigger the planetesimals grew, as the largest swallowed the smaller. Eventually a few dozen big balls of rock, each the size of a small planet, acted as giant vacuum cleaners, sweeping swaths of the Solar System clean of most of its dust and gas as they coalesced and settled into near-circular orbital paths. Where an object ultimately wound up depended in large measure on its mass.

Assembling the Solar System

The Sun, which has the lion's share of the Solar System's mass, dominates everything. Ours is not a particularly massive star system, and so the Sun is a modest sort of star—a good thing for a nearby living planet. Paradoxically, the more massive a star, the shorter its lifetime. The greater interior temperatures and pressures of big stars push nuclear fusion reactions faster and faster. So a star ten times the mass of our Sun might last less than a tenth as long—several hundred million years at most, barely enough time for an orbiting planet to get life started before the star explodes in a killer supernova. Conversely, a red dwarf star a tenth the Sun's mass will last more than ten times longer—one hundred billion years or more—though the energy output of such a weak star might not prove as life-sustaining as our radiant yellow benefactor.

Our middling-size Sun has struck a happy medium: not too large and short-lived, not too small and cool. And its projected nine or ten billion years of reliable hydrogen burning mean there's been plenty of time for life to get going, and there's plenty more for it to continue to evolve. True, in another four or five billion years, the Sun will run out of hydrogen in its core and will have to switch to helium burning. In the process, it will swell to a much less benign red giant star more than one hundred times its present diameter, engulfing poor little Mercury, first scalding and then swallowing Venus, and making things pretty unpleasant on Earth as well. Nevertheless, even after 4.5 billion years, we still have plenty of time before the Sun enters its crotchety old age and life on Earth becomes problematic.

Our Solar System has another important benefit for a living planet. Unlike most others, ours is a single-star system. Astronomers using powerful telescopes have found that about two of every three stars we see in the night sky are actually binaries—systems in which two stars orbit each other in a dance about a common gravitational center. As those stars formed, hydrogen accumulated at two separate places to form big gas balls.

If our nebula had been a bit swirlier, with more angular momentum and consequently more mass out in the region of Jupiter, our Solar System would likely have been a double-star system as well. The Sun would have been smaller, and Jupiter, instead of being a big hydrogen-rich planet, would have grown to become a small hydrogen-rich star. Perhaps life would have thrived betwixt such a polarity. Perhaps an extra star would have provided an extra source of life-sustaining energy. But the gravitational dynamics of two stars can be tricky, and Earth might have ended up a world hostile to life with an eccentric orbit, wobbly spin, and wild climatic swings as two strong gravitational attractors pulled it this way and that.

As it is, the gas giant planets are rather well behaved, with modest size and near-circular orbits around the Sun. Jupiter, the largest of the cast, weighs in at just under a thousandth of the Sun's mass. That's large enough to exert significant control on its planetary neighbors; thanks to Jupiter's disruptive gravitational field, the planetesimals that make up the Asteroid Belt never coalesced into a single planet. But Jupiter is not nearly large enough to trigger nuclear fusion reactions in its own core—the defining difference between stars and planets. The more distant ringed planet Saturn and the even more remote and frigid Uranus and Neptune are smaller still.

Nevertheless, all these gas giant planets were large enough to capture their own gravitationally bound disks of debris, like little solar systems within the Solar System. Consequently, all four outer planets have suites of fascinating moons, including some relatively small asteroids that were attracted and then held in orbit by the giants' gravity. Other moons, some almost as large as the four inner planets and with dynamic geological processes of their own, formed more or less in place from leftover dust and gas—the debris of planetary construction. In fact, the most active object in the Solar System is Jupiter's moon Io, which is so close to the big planet that it completes a full orbit once every forty-one hours. Epic tidal forces constantly stress the moon's 2,260-mile diameter and power a half-dozen volcanoes with sulfur plumes that extend more than a hundred miles from the surface, unlike anything else in the Solar System. Equally intriguing are Europa and Ganymede, big moons roughly the size of Mercury and composed of near-equal proportions of water and rock. Both of these big moons are kept warm inside by Jupiter's incessant tidal forces. Consequently, they both have deep, encircling, ice-covered oceans—places targeted by NASA, in its ongoing search for life on other worlds.

Saturn, the next planet out from the Sun, is endowed with almost two dozen moons, not to mention a glorious ring system dominated by small bits of brilliantly reflective water ice. Most of Saturn's moons are relatively small, some captured asteroids and others formed from Saturn's gassy leftovers; but its largest moon, Titan, is bigger than

Mercury and enshrouded in a thick orange atmosphere. Thanks to the European Space Agency's Huygens lander, which touched down on January 14, 2005, we have close-up views of Titan's dynamic surface. A branching network of rivers and streams feeds frigid lakes of liquid hydrocarbons; the dense, colorful, and turbulent atmosphere is laced with organic molecules. Titan is yet another world worth exploring for signs of life.

The most distant gas giant planets, Uranus and Neptune, are no less well endowed with interesting moons. Most show signs of water ice, organic molecules, and ongoing dynamic activity. Neptune's big moon Triton even has a nitrogen-rich atmosphere. And both Uranus and Neptune have their own complex ring systems, though apparently they are composed of automobile-size chunks of dark carbon-rich material, quite unlike the luminous particles that constitute the icy rings of Saturn.

Rocky Worlds

Closer to home, gravity also held sway. With most of the hydrogen and helium blown outward to the realm of the gas giants after ignition of the Sun, the inner Solar System had much less mass to play with, and most of that consisted of hard rocks—the stuff of chondrite and achondrite meteorites. Mercury, the smallest and driest rocky planet, formed closest to the Sun. A hostile scorched world, this innermost planet appears to be dead and battered: its billions of years of intensely cratered surface is preserved under an airless sky. If you are ever asked to name objects in the Solar System where you'd bet against life, Mercury should top your list.

Venus, the next planet out, is Earth's twin in size but radically different in habitability, thanks in large measure to its orbit, almost thirty million miles closer to the Sun. It may have had a modest store of water early in its history, and even a shallow ocean, but subjected to the Sun's heat and solar wind, most Venusian water appears to have boiled off, preventing that world from being wet. Carbon dioxide, the dominant gas in the thick Venusian atmosphere, sealed in the Sun's radiant energy and created a runaway greenhouse effect. Today Venus's average surface temperatures exceed 900 degrees Fahrenheit—hot enough to melt lead.

Mars, one stop out from Earth, is a lot smaller—only a tenth its mass—but it is in many respects the most Earth-like. Like all the rocky planets, Mars has a metal core and a silicate mantle. Like Earth, it has an atmosphere and a lot of water. Its relatively weak gravity cannot easily hold speeding gas molecules in the upper atmosphere, so billions of years have eaten away at both air and water, and yet Mars still holds warm, wet underground reservoirs where life might maintain a tenuous refuge. No wonder most planetary missions have targeted the red planet.

Earth itself, the "third rock from the Sun," is smack in the middle of the habitable "Goldilocks" zone. It's close enough to the Sun, and hot enough, to have relinquished significant amounts of hydrogen and helium to the outer realms of the Solar System, but it's far enough from the Sun, and cool enough, to have held on to most of its water in liquid form. Like the other planets in our Solar System, it formed about 4.5 billion years ago, essentially from colliding chondrites and their subsequent gravitational clumpings into larger and larger planetesimals, over a span of a few million years.

Deep Time

Layered into all the evidence for how the Sun, the Earth, and the rest of our Solar System were born is the concept of immense time spans—4.5 billion years and counting. Americans love to quote the dates of famous events in human history. We celebrate great accomplishments and discoveries, like the Wright brothers' first flight on

December 17, 1903, and the first manned Moon landing on July 20, 1969. We recount days of national loss and tragedy like December 7, 1941, and September 11, 2001. And we remember birthdays: July 4, 1776, and, of course, February 12, 1809 (the coincident birthdays of Charles Darwin and Abraham Lincoln). We trust the validity of these historic moments because an unbroken written and oral record links us to that not-so-distant past.

Geologists also love to quote historic time markers: about 12,500 years ago, when the last great glaciation ended and humans began to settle North America; 65 million years ago, when the dinosaurs and many other creatures became extinct; the Cambrian boundary, at 530 million years ago, when diverse animals with hard shells suddenly appeared; and more than 4.5 billion years ago, when Earth became a planet in orbit about the Sun. But how can we be sure those age estimates are correct? There's no written record of Earth's ancient chronology past a few thousand years, or any informing oral tradition.

Four and a half billion is a number almost beyond reckoning. The current Guinness world record for longevity is held by a French woman who lived to celebrate her 122nd birthday—so humans fall far short of living even for 4.5 billion seconds (about 144 years). All of recorded human history is much less than 4.5 billion minutes. And yet geologists claim that Earth has been around for more than 4.5 billion years.

There's no easy way to comprehend this *deep time*, but I sometimes try by taking long walks. Just south of Annapolis, Maryland, twenty miles of stately, undulating, fossil-packed cliffs flank the western shore of Chesapeake Bay. Walking along the narrow strip of sand between land and water, one can find an abundance of extinct clams and spiraled snails, corals, and sand dollars. Once in a while on a very lucky day, a six-inch serrated shark tooth or six-foot streamlined whale skull turns up. These prized relics tell of a time fifteen million years ago, when the region was warmer and more tropical, like today's Maui, and majestic whales came to calve and monster sharks sixty feet long feasted on the weak. The fossils populate three hundred vertical feet of sediments, laid down over three million years of Earth history. The layers of sand and marl dip ever so gently to the south, so walking along the beach is like a stroll through time. Each stride to the north exposes slightly older layers.

To get a sense of the scale of Earth history, imagine walking back in time, a hundred years per step—every pace equal to more than three human generations. A mile takes you 175,000 years into the past. The twenty miles of Chesapeake cliffs, a hard day's walk to be sure, correspond to more than 3 million years. But to make even a small dent in Earth history, you would have to keep walking at that rate for many weeks. Twenty days of effort at twenty miles a day and a hundred years per step would take you back 70 million years, to just before the mass death of the dinosaurs. Five months of twenty-mile walks would correspond to more than 530 million years, the time of the Cambrian "explosion"—the near-simultaneous emergence of myriad hard-shelled animals. But at a hundred years per footstep, you'd have to walk for almost *three years* to reach the dawn of life, and almost *four years* to arrive at Earth's beginnings.

How can we be sure? Earth scientists have developed numerous lines of evidence that point to an incredibly old Earth—to the reality of deep time. The simplest evidence lies in geological phenomena that produce annual layerings of material; count the layers, count the years. The most dramatic such geological calendars are varve deposits, thin alternating light and dark layers that represent coarser-grained spring sediments and finer winter sediments, respectively. One meticulously documented sequence from glacial lakes in Sweden records 13,527 years of layering, with a new light-dark layer deposited every year. The finely laminated Green River Shale, which is exposed in scenic steep-walled canyons in Wyoming, features continuous vertical sections with more than a million annual layers. Similarly, ice drill cores thousands of feet deep from Antarctica and Greenland reveal more than eight hundred thousand years of accumulation, year by year, snow layer by snow layer. All of these layerings rest atop vastly older rocks.

Measurements of slower geological processes stretch the timescale of Earth history even further. The massive Hawaiian Islands required slow and steady volcanic activity, as successive lava blankets piled up—at least tens of millions of years, based on modern rates of eruption. The Appalachians and other ancient rounded mountain ranges were formed by hundreds of millions of years of gradual erosion, and the barely detectable motions of tectonic plates that have shifted continents and opened oceans operate in cycles of hundreds of millions of years as well.

Physics and astronomy provide evidence for deep time that is no less persuasive. The predictable decay rates of radioactive isotopes of carbon, uranium, potassium, rubidium, and other elements are exceptionally accurate clocks for dating rock-forming events that extend back billions of years to the formation of the Solar System. If you have a collection of one million atoms of a radioactive isotope, half of them will decay over a span of time called the half-life. Leave behind a million atoms of uranium-238, for example, and come back when its half-life of 4.468 billion years has passed, and you'll find only about a half-million atoms of uranium-238 remaining. The rest of the uranium will have decayed to a half-million atoms of other elements, ultimately to stable atoms of lead-206. Wait another 4.468 billion years and only about a quarter-million atoms of uranium will remain. The age determinations of the oldest primitive chondrites—4.566 billion years old—are obtained by this radiometric dating method.

But what of the many billions of years *before* the Solar System? Astrophysicists' measurements of distant galaxies in motion point to a universe that is much older than 4.5 billion years. All galaxies are speeding away from us. Doppler shift data—the so-called redshift— reveal that the more distant galaxies are retreating even faster. Play that cosmic tape backward, and everything converges down to a point about 13.7 billion years ago. That's the Big Bang. The light from some of these most distant objects has been traveling through space for more than 13 billion years.

The data on this point are unassailable. Any claim that Earth's age is ten thousand years or less defies the overwhelming and unambiguous observational evidence from every branch of science. The only alternative is that the cosmos was created ten thousand years ago to look vastly older—a conclusion first expounded by American naturalist Philip Gosse in 1857, in his convoluted treatise *Omphalos* (named after the Greek word for "navel," because motherless Adam was created with a navel, so as to look as if he were born by woman). Gosse cataloged hundreds of pages of evidence for an extremely ancient Earth and then proceeded to describe how God created everything ten thousand years ago *to look much older.*

Some may find comfort in this Creationist loophole of created antiquity, known as prechronism. To astrophysicists' observations of stars and galaxies that are billions of light-years away, prechronists retort that the universe was created with light from those stars and galaxies that was already on its way to Earth. Rocks with ancient ratios of radioactive and daughter isotopes, they argue, were created with just the right mixtures of uranium, lead, potassium, and argon to make them *appear* much older than they really are. If you are of this prechronist persuasion, I suggest you skip ahead to chapter 11, "The Future." Otherwise, allow your imagination to skip back several billion years into the past, when our planet was born.

The birth of Earth 4.5 billion years ago was a drama that has been repeated countless trillions of times throughout the history of the universe. Every star and planet arises in the tenuous near-vacuum of space from gas and dust—individual particles of matter too small to see with the unaided eye, yet so vast in total extent that we can observe immense star-forming clouds halfway across the galaxy. Billions of years ago, gravity served to midwife the Solar System's birth—the Sun emerged as the solitary giant in a litter of planetary runts. Nuclear reactions inflamed the Sun's surface, bathing its planetary neighbors in light and warmth. And so our home took its first halting steps to becoming a living world.

As alien as such epic events might seem, we all experience, every day of our lives, the same cosmic phenomena that led to Earth's formation. The very same elements and atoms that forged Earth also make up our bodies and our dwellings. The same universal force of gravity that assembled the stars and planets from dust and gas, and that forged the elements into stars, also holds us fast to our planetary home. When it comes to the universal laws of physics and chemistry, there is nothing new under the Sun.

The lessons of rocks, stars, and life are equally clear. To understand Earth, you must divorce yourself from the inconsequential temporal or spatial scale of human life. We live on a single tiny world in a cosmos of a hundred billion galaxies, each with a hundred billion stars. Similarly, we live day by day in a cosmos aged hundreds of billions of days. If you seek meaning and purpose in the cosmos, you will not find it in any privileged moment or place tied to human existence. The scales of space and time are too inconceivably large. But a cosmos bound by natural laws that lead inevitably, inexorably to a universe that promises the possibility of knowing itself, as scientific study inherently suggests, is a cosmos that abounds with meaning.

Chapter 2

The Big Thwack

The Formation of the Moon

Earth's Age: 0 to about 50 million years

↜ A central principle of this book is that planetary systems evolve—they change through time. What's more, each new evolutionary stage depends on the prior sequence of stages. Changes are often gradual, taking millions or even billions of years to transform a planet's environment, but sudden violent and irreversible events can forever alter a world in minutes. So it was with Earth.

From countless scattered bits and pieces, Earth formed relatively quickly, in no more than a million years by some estimates. Toward the end of this process a few dozen planetesimals, each several hundred miles in diameter, shared space with the proto-Earth. In a span of about a hundred thousand years, as our planet approached its present size, the final stages of this process occurred in episodes of unfathomable violence. Once every few thousand years, a miniplanet smacked into the proto-Earth and was swallowed whole.

During that turbulent time, Earth was a hot, blackened sphere, punctuated with glowing red cracks, towering volcanic magma fountains, and incessant meteor impacts. Each of the giant impactors smashed into the sphere, blasting vaporized rocks into orbit and disrupting the entire surface into a molten, red-hot rocky slush. Space is cold, however. Following each great impact, Earth's airless surface quickly cooled and blackened again.

Strange Moon

This story of Earth's origins seems to be neat and tidy except for one striking detail: the Moon. It's way too big to ignore and, for much of the past two centuries, has proven exceedingly difficult to explain. Small moons are easy to understand. Phobos and Deimos, the two irregular city-size rocks orbiting Mars, appear to be captured asteroids. The dozens of much larger moons orbiting Jupiter, Saturn, Uranus, and Neptune are teeny by comparison to their hosts—much less than a thousandth of the mass of the planets they circle. The largest moons, formed from unclaimed remnants of the original planet-forming dust and gas, orbit these gas giants like planets in miniature solar systems. Earth's Moon, by contrast, is relatively huge compared with the planet it orbits: it has more than a quarter of Earth's diameter and about one-eightieth of its mass. Where did such an anomaly come from?

The historical sciences, especially Earth and planetary sciences, depend on creative storytelling (albeit stories that more or less conform to the facts). If more than one story seems to fit the observations, then geologists adopt a cautious stance known as "multiple working hypotheses"—a strategy familiar to anyone who enjoys detective novels.

Prior to the historic Apollo Moon landings beginning in 1969, when pristine Moon rocks were recovered and careful geophysical measurements of the Moon's interior could begin, three prime suspects stood out in The Case of the Massive Moon. The first widely accepted scientific hypothesis was the fission theory, proposed in 1878 by George Howard Darwin (who is far less famous than his naturalist father, Charles). In George Darwin's scenario, the primordial molten Earth was spinning on its axis so rapidly that it stretched and elongated until it flung off a glob of magma from the surface into orbit (with a little help from the Sun's gravitational pull). The Moon, in this model, is an Earth bud broken free. In one imaginative variant of this dramatic tale, the Pacific Ocean basin remains as a telltale mark—Mother Earth's birthing scar.

A second competing idea, the capture theory, viewed the Moon as a separately formed, smaller planetesimal occupying more or less the same zip code as Earth, in the emerging Solar System. At some point, the two bodies passed close enough to each other that larger Earth captured smaller Moon, swinging it into a looping orbit that has gradually settled down. That greedy gravitational mechanism seemed to work well enough for the smaller rocky moons of Mars, so why not for Earth?

The third hypothesis, the co-accretion theory, posited that the Moon formed more or less in its present location from a large cloud of leftover debris that remained in orbit around the Earth. This plausible idea mimics what we know about the Sun and its planets, as well as the gas giant planets and their moons. It's a common theme, seen over and over again in the Solar System: smaller objects accrete from clouds of dust, gas, and rocks around larger objects.

Three competing hypotheses; which one is correct? Inquiring minds had to await data from the Moon rocks—more than 840 pounds of samples from six Apollo landing sites.

Touchdown on the Moon

The Apollo Moon missions transformed planetary science in many ways. Sure, they were an unrivaled showcase for American technological prowess and bravado. Undoubtedly, they provided a tremendous boost to the military-industrial complex. And they inspired countless innovations, from minicomputers to polymers to Tang, providing an economic driver that may well have paid for the $20 billion worth of missions many times over. It's not surprising that national pride and the race for the "high ground," not lunar science, were the primary incentives for those costly and dangerous early Moon missions.

Even so, it would be hard to overstate the impact of the Apollo missions and their treasure trove of Moon rocks on my generation of Earth scientists. For all of human history, the Moon was tantalizingly close, just under a quarter-million miles away. On a clear summer's evening, as the reddened full Moon rises, you feel as if you could just reach out and touch it. But we had no samples—nothing to tell us for sure of what the Moon was made, nor when, nor where. With the first batch of lunar samples, we could, for the first time in human history, literally touch the Moon (as can any visitor to the Smithsonian today).

My literal first breath of lunar samples came in the winter of 1969–70, during my senior year at MIT, less than half a year after Apollo 11's historic mission. The stage had been set a few months earlier, on July 24, 1969, when the first humans to walk on the Moon returned to Earth. In those early days of lunar exploration, concerns of contamination by alien microbes dictated strict quarantine policies for astronauts and their samples. So as soon as their module splashed down in the Pacific near Hawaii, as soon as Neil Armstrong, Buzz Aldrin, and Mike Collins were retrieved by the USS *Hornet,* they and the forty-five priceless pounds of rocks and soil they had brought home from space were hermetically sealed in NASA's Mobile Quarantine Facility. From Hawaii, they were all shipped to Houston, to the new Lunar Receiving Laboratory, where the space explorers and their precious samples were confined for almost three weeks, in case something really nasty had accompanied them back to Earth.

Apollo missions followed fast upon one another in the next three years. The Apollo 12 lunar module *Intrepid,* with astronauts Charles

Conrad, Jr., and Alan Bean, touched down on November 19, 1969, and a week later returned with about seventy pounds of Moon rocks and soil, which were whisked into the Houston quarantine facilities. By good fortune, my thesis adviser, the brilliant and ebullient David Wones, was a member of the Apollo 12 Lunar Sample Preliminary Investigation Team. That small band of scientists had the glorious adventure of scrutinizing the second precious lunar sample haul with a state-of-the-art arsenal of analytical machines. Dave's expertise was igneous petrology—the study of the origin of rocks that form from magma. All of the Apollo 11 and 12 Moon rocks were igneous in origin, so he was in geologist heaven.

In some ways, it was hard duty, being locked up for the better part of a month with a few other intense scientists, under pressure to gather unassailable data on some of the most costly and significant rock samples ever collected. But it was also incredibly exciting to be among the first humans to handle rocks and soil from another world—the space matter that would once and for all tell us the origin of the Moon.

My first glimpse of the Moon up close came on Dave's return to MIT. I recall the elevator door opening on the twelfth floor of the Green Building. There was Dave, of modest stature and bespectacled, flanked by two beefy, uniformed, gun-toting federal agents. They were guarding the Moon samples, which at that point would have been worth millions on the collectors' market. Every milligram had to be accounted for. Dave looked tired and nervous; he had been away for a long time, he was under constant scrutiny, and he still had a job to do.

When the subject of lunar samples comes up, most people immediately think of Moon rocks, perhaps something chunky that you can hold in your hand. But a significant portion of the Apollo material was lunar soil, or regolith. The finest-grained fraction of regolith is pulverized rock in fragments so small that you can't resolve them in a microscope—the consequence of a battery of cosmic insults, from mighty asteroids to the incessant solar wind. This ultrapowder has strange properties, most notably that it sticks to everything it touches, like Xerox toner. Dave's task was to transfer some of this powder from a vial about the size of a C battery into three or four smaller vials, about the size of AAA batteries, for distribution to nearby laboratories.

It sounds easy. Dump the powder from the big vial onto a three-inch-square piece of glassy-surfaced powder paper. Gently scoop small amounts into the smaller vials. Dave had performed similar operations hundreds of times, and it shouldn't have taken more than a minute. But the stakes were higher here. Humorless guards were standing to either side; a small cohort of students was hovering, too. So Dave's hand shook a bit as he tipped the big vial. The sticky powder clung to the glass sides and didn't want to come out. He tapped it with his index finger. Nothing. Tapped again.

Then all of a sudden all the Moon dust—really only a little pile the size of a Hershey kiss, but it seemed like a lot under the circumstances—came shlumping out all at once, and then poof! Dust flew up, coated Dave's fingers, and spilled over the edge of the powder paper onto the table. We all must have breathed in some of the finest airborne particles. No one said a word.

It wasn't a real disaster, as almost nothing was lost, and the powder did eventually get transferred, and the feds did eventually leave to hand off the aliquots to other labs. In retrospect, we all thought it was pretty funny. And in a couple of days, above the lab bench where the transfer had been completed, we neatly framed that three-inch square of powder paper with a near-perfect imprint of Dave Wones's left index finger in Moon dust.

Four more Apollo Moon landings followed, culminating in December 1972 with Apollo 17 and the return of more than 240 pounds of samples from the Taurus-Littrow Valley, a region of suspected lunar volcanism. That was the last mission; no one has gone back in four decades. Nevertheless, the Moon rocks, meticulously curated in sterile vaults at the Lunar Sample Building at NASA's Johnson Space Center in Houston (with a secure backup collection at Brooks Air Force Base in San Antonio, Texas), continue to provide an amazing wealth of opportunities for researchers.

A few years after the last Apollo mission, those samples provided me my first real job, as a postdoctoral fellow with the Geophysical Laboratory of the Carnegie Institution. My task was to examine piles of "fines" from Apollo 12, Apollo 17, and LUNA 20 (one of three unmanned Soviet missions that had returned about a third of a pound of lunar samples). Interspersed throughout the fine dust of lunar soil are lots of silt- and sand-size grains, and my exacting job was to scan thousands of these grains, bit by bit. I spent hours at a microscope, peering at gorgeous little green and red crystals and tiny golden spheres of colorful glass—the remains of violently blasted rocks that had been subjected to billions of years of meteorite bombardment.

Once I'd isolated a few dozen promising specks, I subjected each unusual grain to three kinds of analysis. The first was single-crystal X-ray diffraction, to tell what kind of crystal I was dealing with. Most of my studies focused on the common minerals olivine, pyroxene, and spinel. If I found a good crystal, I'd carefully orient the grain and measure its optical absorption spectrum (the way it soaked up different wavelengths of light). Green olivine crystals, for example, typically absorb red wavelengths; red spinel crystals, by contrast, absorb more in the green wavelengths. I also measured spectra of any unusual glass beads, keeping an eye out for telltale bumps and wiggles in the absorption spectrum that pointed to rarer elements—chromium and titanium, for example. Discovery of a small peak at 625 nanometers, a slight absorption of red-orange wavelengths characteristic of the element chromium as it occurs on the Moon but quite different from chromium on Earth, was a memorable "eureka" moment.

Finally, after the X-ray and optical work was done, I used a fancy analytical machine called an electron microprobe to determine the exact ratios of elements in my samples. Time and time again I confirmed what others had found: the Moon's surface minerals, while similar in major elements to those on Earth, are rather different in detail. They have more titanium; the chromium is different, too.

These and other clues from the Apollo rocks placed severe constraints on the various theories of how the Moon came to be. For one thing, it turned out that the Moon differs dramatically from Earth, in that its density is much lower; it doesn't have a big, dense iron metal core. Earth's core holds almost a third of its mass, but the Moon's tiny core is less than 3 percent of its mass. Second, Moon rocks contain almost no traces of the most volatile elements—those that tend to vaporize the moment things get warm. The nitrogen, carbon, sulfur, and hydrogen so common at Earth's surface are missing from Moon dust. This deficiency means that unlike Earth, which is covered in liquid water and whose soils contain abundant water-rich minerals such as clays and micas, no water-bearing minerals of any kind have been brought back from the Apollo missions. Something must have blasted or baked the Moon to remove those volatiles, for the Moon's surface is now an unforgivingly dry place.

The third key finding of the Apollo missions is based on the element oxygen, or more specifically the distribution of its isotopes. Each chemical element is defined by the number of positively charged protons in its nucleus. The number is unique—*oxygen* is just another name for "atom with eight protons." Atomic nuclei also hold a second kind of particle, the electrically neutral neutron. More than 99.7 percent of oxygen atoms in the universe have eight neutrons (eight protons plus eight neutrons yields an isotope called oxygen-16), while the rarer isotopes with nine or ten neutrons (oxygen-17 and oxygen-18, respectively) are present at a fraction of a percent.

Oxygen-16, oxygen-17, and oxygen-18 are virtually identical in their chemical behavior—you could breathe any mixture and wouldn't notice a difference—but they do have different masses. Oxygen-18 is heavier than oxygen-16. Consequently, anytime an oxygen-containing compound changes state from a solid to a liquid, or from a liquid to a gas, the less massive oxygen-16 can make the move more easily. In the turbulent nascent Solar System, such changes of state were commonplace, and they led to shifting amounts of oxygen isotopes. It turned out that the ratio of oxygen-16 to oxygen-18 varies from planet to planet and is very sensitive to the planet's distance from the Sun when it formed. The Apollo rocks revealed that the Moon's oxygen isotope ratio is virtually identical to Earth's. In other words, Earth and the Moon must have formed at about the same distance from the Sun.

So where did these discoveries leave the three competing Moon-forming hypotheses? The co-accretion theory was in trouble from the start. If the Moon formed from Earth leftovers, then it should have a similar average composition. True, the Moon and Earth do match up in terms of oxygen isotopes, but the co-accretion theory can't explain the large differences in iron and volatiles. The Moon's bulk composition is just too different for it to have formed from the same stuff as Earth.

The compositional disparities also posed insurmountable problems for the capture theory. Theoretical models of planetary motion suggest that any captured planetesimal must have formed in the solar nebula at more or less the same distance from the Sun as Earth, and so it should have more or less the same average composition. The Moon doesn't. Of course, a Moon-size object might have formed in some other district of the solar nebula and subsequently adopted an Earth-crossing orbit, but computer models of orbital dynamics require that such a Moon would have had a high velocity relative to Earth, making such a capture scenario all but impossible.

That leaves George Howard Darwin's fission theory. It can successfully explain the similar oxygen isotope compositions (Earth and Moon are one system) and the iron difference (Earth's core had already formed; the Moon-forming blob was a chunk of Earth's already differentiated, iron-poor mantle). It beautifully accommodates the fact that one side of the Moon always faces Earth: Earth's rotation and the Moon's orbit follow the same spinning motion around the Earth's axis—the same sense of turning. But a big problem remains: where are the missing volatiles, now absent from the Moon?

The laws of physics also get in the way of the fission theory. At about the time of the Apollo missions, computer modeling of planet formation had progressed to a point that theorists could study the dynamics of a rapidly spinning Earth-size ball of magma with confidence. Simply put, fission can't work. Earth's gravity is much too strong to allow a big blob of molten rock to be tossed outward into orbit. In fact, a molten Earth would have had to be spinning on its axis at an incredible rate, about once every hour, for it to fling off a Moon-size glob. The Earth-Moon system simply does not have enough angular momentum for that to happen.

Bottom line: none of the three prevailing theories of Moon formation fits the data after the Apollo missions. There must be another explanation.

The Testimony of the Moon Rocks

Planetary scientists are nothing if not good storytellers. Observations from Apollo disproved all three of their pre-1969 hypotheses about the Moon's formation, but it didn't take them long to come up with a new idea from the indisputable facts. New compositional clues from

Apollo provided one key: the Moon more or less looks like Earth. It has the same oxygen isotope composition and most of the same major elements, though it has way too little iron or volatiles. That compositional data had to be integrated with the orbital clues we'd known for thousands of years: the Moon orbits Earth in the same plane and in the same direction as the other planets around the Sun. Earth does have a niggling 23-degree tilt in its rotation axis (that's what causes the seasons). And one side of the Moon always faces us.

Earlier models of lunar formation tended to ignore orbital clues beyond the Earth-Moon system, including some striking exceptions to the general patterns of our Solar System. For one thing, Venus rotates on its axis backward compared with all the other planets. That may not seem significant, but Venus is almost as large as Earth—and it's *rotating the wrong way*! Even stranger is massive Uranus, the third largest planet, whose rotation axis is sideways, so it seems to kind of roll along its orbit around the Sun. The moons of other planets, too, have oddities. Neptune's largest moon, Triton, which is comparable in size to Earth's Moon, orbits at a steep angle to the planet's rotation and *in the opposite direction of the rest of the Solar System*.

The culture of science has an odd aspect—one that may be off-putting to those not in the game. On the one hand, we come up with neat theories that package lots of odd facts. So the fact that all the planets and moons orbit the Sun in the same direction, in the same plane, points to a common origin from a single swirling nebula. But then we find exceptions to the rule, and those exceptions are set aside as curious anomalies. Venus rotates the wrong way? Triton orbits the wrong way? No problem. Their aberrations are incidental to the larger scheme.

The same kind of situation complicates many public debates, like that over global warming. Many scientists predict that altered atmospheric conditions will raise the average global temperature by several degrees. But such changes can also cause extreme weather, which may mean worse snowstorms in the southern United States. Global warming may alter ocean currents like the Gulf Stream and ultimately turn northern Europe into a much colder Siberian-type icebox. Anomalies like this fuel the global warming naysayers: scientists say the world is getting hotter, but you've just suffered through the biggest snowstorm in your region's history. How should you respond? A judicious response is that nature is amazing—rich, varied, complex, and intricately interconnected, with a messy, long history. Anomalies, whether in planetary orbits or North American weather, are not just inconvenient details to brush aside: they are the very essence of understanding what really happened—how things really work. We develop grand and general models of how nature works, and then we use the odd details to refine the original imperfect model (or if the exceptions overwhelm the rule, we regroup around a new model). That's why good scientists revel in anomalies. If we understood everything, if we could predict everything, there'd be no point in getting up in the morning and heading to the lab.

In the case of the origin of Earth's Moon, those exceptions to the systematic trends—those niggling orbital anomalies—led to the concept of the "Big Splash" or "Big Thwack" model, which arose in the mid-1970s. The original series of related but poorly constrained hypotheses coalesced into conventional wisdom at a pivotal 1984 conference in Hawaii, where planetary formation experts gathered to weigh all their options. In such a heady environment, Ockham's razor—the demand that the simplest solution to a problem consistent with the facts is likely to be correct—prevails. The Big Thwack fit the bill.

To understand this radical idea, think back more than 4.5 billion years, to the time when the planets had just formed from all those smaller competing planetesimals. As Earth grew close to its present diameter of eight thousand miles, it swallowed up almost all the remaining proximate bodies in a succession of huge impacts. Those penultimate collisions with objects many hundreds of miles across would have been spectacular, but they had little effect on Earth, the much more massive proto-planet.

But not all impacts are equal. In Earth history, one single event—one day more memorable than any other—stands out. About 4.5 billion years ago, when the Solar System was about 50 million years old, the black proto-Earth and a slightly smaller planet-size competitor were jockeying for the same narrow band of Solar System real estate. The smaller would-be planet (dubbed Theia, after the Titan goddess who gave birth to the Moon) was worthy of planetary status—perhaps two to three times the size of Mars (or roughly a third of Earth's mass). A rule of astrophysics is that no two planets can share the same orbit. Eventually they will collide, and the larger planet always wins. So it was with Earth and Theia.

Increasingly vivid computer simulations provide the principal method by which scientists attempt to understand what might have happened. A big collision is governed by the laws of physics, so one can run thousands of simulations with all sorts of initial conditions to see if a Moon results. The answer is intimately tied to the starting parameters: the mass and composition of the proto-Earth, the mass and composition of Theia, their relative velocities, and the angle and accuracy of the blow. Most combinations simply do not work; no Moon forms. But a few models are surprisingly successful and produce an Earth-Moon system rather like the one we see today.

In one oft-described version, the impact occurs as a solid side-swipe—big Theia smashes the bigger Earth slightly off center. Seen from space, the event plays out in slow motion. At the moment of contact, the two worlds seem at first to gently kiss. Then over the next four or five minutes, Theia is smooshed, like a ball of soft dough hitting the floor, without much effect on Earth. Ten minutes later Theia is pretty much squashed, while Earth begins to deform out of roundness. Half an hour into the collision, Theia is simply obliterated, while the injured Earth is no longer a symmetrical sphere. Superhot rock has been vaporized, blasting out in luminous streams from the gaping wound and obscuring the disrupted worlds.

Another widely cited scenario, first proposed in the 1970s and refined over the next two decades, was developed by theorist Alastair Cameron of the Harvard-Smithsonian Center for Astrophysics. In his intriguing scheme, Theia was roughly 40 percent the mass of the proto-Earth. Again, an off-center impact occurred, but in this version, Theia more or less bumped against Earth and bounced off as an elongated blob, then was pulled back in for the coup de grâce—a second thwack, in which Theia disappeared forever.

In either case, the catastrophe annihilated Theia, which simply vaporized into an immense incandescent cloud, tens of thousands of degrees hot, surrounding Earth. Theia had done its share of damage as well. A significant chunk of Earth's crust and mantle also vaporized and blasted outward to mix with Theia's scattered remnants. Some material escaped to deep space, but most of the savaged remains were retained in orbit by Earth's unyielding gravitational grip. From this roiling cloud, dense metal from the cores of both worlds commingled and cooled back into liquid, sinking to form a new, larger core for Earth. Mantle materials also mixed and vaporized, forming a hellishly hot globe-encircling cloud of vaporized rock. For a violent time of days or weeks, Earth experienced an incessant rain of orange-hot silicate droplets, which merged with a shoreless, red-glowing magma ocean. Ultimately Earth seized much of what had been Theia and thus emerged a more massive planet.

Not all of Theia was captured, however. Higher up in space, Earth became encircled by a vast accumulation of rocky collisional debris, mostly an intimate mixture of the two planetary mantles. Cooling rocky droplets stuck together, with bigger chunks sweeping up the smaller. In a sort of instant replay of the gravitational clumping that originally formed the planets, the Moon coalesced rapidly and may have achieved more or less its present size in a few years.

The physics of planet formation dictate where the Moon could have formed. Every massive object has an invisible surrounding sphere, called the Roche limit, inside of which gravitational forces are too great for a satellite to form. That's why Saturn has immense rings but no moons within about fifty thousand miles of its surface. Saturn's gravity prevents those icy particles from coalescing to form a moon.

Calculated from the center of a rotating object, Earth's Roche limit is about 11,000 miles, or roughly 7,000 miles up from the surface. Accordingly, models of Moon formation locate the new satellite at a safe distance of about 15,000 miles up, where it could grow in an orderly fashion by sweeping up most of the scattered bits and pieces from the Big Thwack. And so, perhaps 4.5 billion years ago by most estimates, the Moon was born. Earth found itself with a companion, formed in large part from pieces of itself.

Scientists quickly embraced the Big Thwack theory because it explains all the major clues better than any other model. The Moon lacks an iron core because most of Theia's iron wound up inside Earth. The Moon lacks volatiles because Theia's volatiles were blasted away during the impact. One side of the Moon always faces Earth because the angular momentum of Earth and Theia were coupled into one spinning system.

The Big Thwack also helps to explain Earth's anomalous axial tilt of about 23 degrees—a factor not well handled by any of the previous scenarios. The impact of Theia literally tipped Earth onto its side. Indeed, the realization that a giant impact formed the Moon has led to speculation about other planetary anomalies in our Solar System. Perhaps late Big Thwack events of one kind or another are common, even necessary. Perhaps that explains why Venus rotates the wrong way on its axis and why it lost so much of its water. Perhaps a late giant impact caused Uranus to rotate on its side.

A Different Sky

The Moon's formation was a pivotal moment in Earth history, with far-reaching consequences that are utterly amazing and only just now coming into focus. The Moon of 4.5 billion years ago was not the romantic, silvery disk we see today. Long ago it was a far more looming, dominant, and unimaginably destructive influence on Earth's near-surface environment.

It all boils down to one amazing fact: the Moon formed only 15,000 miles from Earth's surface—not much farther than a plane flight from Washington, D.C., to Melbourne, Australia. Today, by contrast, the Moon is 239,000 miles away. At first blush, it seems utterly implausible that a giant Moon could just drift away from Earth like that, but measurements don't lie. Apollo astronauts left shiny mirrors on the lunar surface. Laser beams from Earth bounce off the mirrors and return to Earth to provide distance measurements accurate to within a tiny fraction of an inch. Every year since the early 1970s, the Moon has moved farther away: 3.82 centimeters per year, about one and a half inches per year on average. That doesn't sound like much, but it adds up over time—to a mile farther away every forty thousand years at present rates. Play the tape backward, and it points to a radically different situation 4.5 billion years ago.

For one thing, the Moon *looked* totally different. At 15,000 miles distant, the 2,160-mile-diameter Moon would have appeared gigantic, like nothing we've ever seen. It spanned almost 8 degrees of arc in the sky—roughly sixteen times the apparent diameter of the Sun—and blocked more than 250 times as much of the firmament as the Moon does today.

And that's not all. The early Moon was a violent body of intense volcanism, quite unlike the static silvery-gray object we see now. Its surface would have appeared black, with glowing red magma-filled cracks and volcanic basins easily visible from Earth. The primordial full Moon was equally dramatic, the surface reflecting hundreds of times more sunlight than in our modern era. You could easily read a book under its brilliant illumination, but astronomical observations would have been futile. No stars or planets would have been visible against the young Moon's intense glare.

Adding to the drama was how fast things moved then. In space, there is no friction, so spinning objects just keep spinning for billions of years. Spinning systems like the Moon plus Earth possess the property of angular momentum—a quantity measured by a combination of two familiar circular motions. First is Earth's rotation about its axis; the faster Earth rotates, the more angular momentum it has. The Moon's angular momentum, by contrast, depends primarily on how far away and how fast it orbits around Earth. Its own rotation is not a significant part of the equation.

The total angular momentum of Earth's rotation plus the Moon's orbit hasn't changed significantly over the last several billion years, but the relative importance of those two motions has changed a lot. Today almost all the angular momentum of the Earth-Moon system is tied up in the orbiting Moon, with its 239,000-mile distance and twenty-nine-day orbital period. The more massive central Earth, with its leisurely twenty-four-hour day, has only a tiny fraction of the Moon's angular momentum. (By the same token, the distant gas giant planets carry almost all the Solar System's angular momentum, even though the central Sun has 99.9 percent of the mass.)

But 4.5 billion years ago, things were very different. With the Moon only 15,000 miles away, everything was turning ridiculously fast, like the ice-skater who's pulled in her arms to speed up her spin. For one thing, Earth rotated on its axis once every five hours. It still took a full year (about 8,766 hours) to go around the Sun; that time hasn't changed much in the history of the Solar System. But there were more than 1,750 short days per year, and the Sun rose every five hours!

Such an estimate seems bizarre and untestable, but at least a couple of direct measurements confirm this idea of shorter ancient days. Coral reefs are one compelling form of evidence. Some species of coral display exceedingly fine-scale growth lines that record both subtle daily and more obvious annual cycles. As expected, modern corals show about 365 daily lines for every year of growth. But ancient fossil corals from the Devonian Period, about four hundred million years ago, display more than four hundred daily lines per year, pointing to a faster rotation rate. Days were only about twenty-two hours long back then, when the Moon was perhaps ten thousand miles closer to Earth.

A second, complementary measurement rests on the euphonic phenomenon of tidal rhythmites, which are finely layered sediments that reveal the daily, lunar, and yearly cycles of the tides. Exacting microscopic studies of tidal rhythmites from nine-hundred-million-year-old rocks at Big Cottonwood Canyon, Utah, point to a world when Earth days were only 18.9 hours long, when there may have been 464 days—464 sunrises and sunsets—every year. The calculated Earth-Moon distance of 218,000 miles at that time implies a recession rate very similar to that of modern times: 3.91 centimeters per year, slightly more than one and a half inches annually.

Loony World

No direct evidence yet documents Earth's tidal cycles more than a billion years ago, but we can be confident that 4.5 billion years ago things were a lot wilder. Not only did Earth have five-hour days, but the nearby Moon was much, much faster in its close orbit, as well. The Moon took only eighty-four hours—three and a half modern days—to go around Earth. With Earth spinning so fast and the Moon orbiting so fast, the familiar cycle of new Moon, waxing Moon, full Moon, and waning Moon played out in frenetic fast-forward: every few five-hour days saw a new lunar phase.

Lots of consequences follow from this truth, some less benign than others. With such a big lunar obstruction in the sky and such rapid orbital motions, eclipses would have been frequent events. A total solar eclipse would have occurred every eighty-four hours at virtually every new Moon, when the Moon was positioned between Earth and the Sun. For some few minutes, sunlight would have been completely blocked, while the stars and planets suddenly popped out against a black sky, and the Moon's fiery volcanoes and magma oceans stood out starkly red against the black lunar disk. Total lunar eclipses occurred regularly as well, almost every forty-two hours later, like clockwork. During every full Moon, when Earth lies right between the Sun and the Moon, Earth's big shadow would have completely obscured the giant face of the bright shining Moon. Once again the stars and planets would have suddenly popped out against a black sky, as the Moon's volcanoes put on their ruddy show.

Monster tides were a far more violent consequence of the Moon's initial proximity. Had both Earth and the Moon been perfectly rigid solid bodies, they would appear today much as they did 4.5 billion years ago: 15,000 miles apart with rapid rotational and orbital motions and frequent eclipses. But Earth and the Moon are not rigid. Their rocks can flex and bend; especially when molten, they swell and recede with the tides. The young Moon, at a distance of 15,000 miles, exerted tremendous tidal forces on Earth's rocks, even as Earth exerted an equal and opposite gravitational force on the largely molten lunar landscape. It's difficult to imagine the immense magma tides that resulted. Every few hours Earth's largely molten rocky surface may have bulged a mile or more outward toward the Moon, generating tremendous internal friction, adding more heat and thus keeping the surface molten far longer than on an isolated planet. And Earth's gravity returned the favor, bulging the Earth-facing side of the Moon outward, deforming our satellite out of perfect roundness.

These epic tidal disruptions lie at the heart of why the Moon keeps moving away from Earth. How does a 2,160-mile-wide object drift from a mere 15,000 miles to 239,000 miles away? The answer is found in the conservation of angular momentum—the constant sum of Earth's angular momentum plus the Moon's angular momentum. The laws of physics dictate that whatever angular momentum the Earth-Moon system had at its origin, it must still possess in large measure today.

Four and a half billion years ago, a great tidal bulge swept around planet Earth every few hours. But because Earth's surface revolved around its axis faster (every five hours) than the Moon orbited around that same axis (every eighty-four hours), the tidal bulge with its extra mass was always in the lead, constantly pulling on the Moon with the force of gravity, transferring angular momentum from Earth to the Moon with every orbit. The immutable laws of planetary motion, first proposed about four hundred years ago by the German mathematician Johannes Kepler, state that the more angular momentum a satellite possesses, the farther away it has to be from its central planet. With every orbit, the Moon inexorably receded from Earth.

At the same time that Earth's tidal bulge pulled on the Moon, the tidally deformed Moon pulled back on Earth's massive bulge with equal and opposite gravitational force, thus making Earth rotate more slowly on its axis with every rotation. That's where conservation of angular momentum comes in. The faster the Moon orbited, the farther it had to be from Earth and the more angular momentum it picked up. To compensate, Earth had to rotate ever more slowly on its axis to conserve the total angular momentum of the Earth-Moon system—again, think of the figure skater, stretching out her arms once more and slowing her spin. Over the span of 4.5 billion years, Earth's rotation has slowed from once every five hours to once every twenty-four hours, while the Moon has moved farther away and picked up a lot of angular momentum in the process.

Not every planet-moon system has to follow this storyline. If the planet rotates on its axis more slowly than its moon orbits, then an inexorable braking effect ensues. Tidal bulges on the planet will trail behind; the moon will slow down with each orbit and fall ever closer to its doom. Eventually the moon will spiral into the planet and be swallowed up, in yet another variation on the Big Thwack theme. Perhaps that's why Venus, with its wrong-way retrograde rotation, doesn't have a moon. Perhaps such a cataclysmic demise of a once-orbiting moon explains why Venus lost its water and is now a hostile, scalding, lifeless world.

Early in the history of the Earth-Moon system, these exchanges of angular momentum from the slowing Earth to the accelerating Moon were vastly greater than today. In the first centuries after the Moon's formation, both bodies were girdled by turbulent magma oceans that could flex and deform. The giant magma tides on Earth, and similar magma bulging on the Moon, probably caused the Moon to recede by tens or hundreds of feet per year, even as Earth's rotation steadily slowed down from its initial frenetic pace. But these enormous land tides could not have lasted for long. As the Earth-Moon distance increased, the tidal forces decreased even more: a doubling of the distance cut the force of gravity by a factor of four. A tripling of the distance and gravitational forces were but a ninth of their former strength.

Repeated tidal stressing delayed but could not stop the solidification of worlds. Within a few million years of the Big Thwack, the surfaces of both Earth and the Moon were paved with hard black rock. Land tides—the deformation of solid rock—were still not trivial in those early days, but they were nothing like the mighty daily swellings of the magma sea that preceded them.

The Moon remains a luminous reminder that the cosmos is a place of intertwined creation and destruction. Even today we are not immune to catastrophic cosmic insults: killer asteroids and comets still cross Earth's orbit from time to time. Millions of years ago one big rock killed the dinosaurs; millions of years from now, other big rocks will inevitably find their mark. If human survival is our greatest collective imperative as a species, then we would do well to keep watching the skies, for our nearest cosmic neighbor offers mute testimony: while change is usually gradual and benign, there can be really bad days.

Chapter 3

Black Earth

The First Basalt Crust

Earth's Age: 50 to 100 million years

ᖇᖇ Earth has suffered more than a few transformative events in its long history. The Big Thwack was surely the most disruptive and, in the consequent formation of the Moon, had perhaps the most far-reaching effects. But such an outcome—a large solitary Moon orbiting a planet full of volatiles—is by no means an inevitable outcome of the laws of chemistry and physics. Had details of that ancient interaction between Earth and Theia unfolded with only slight variations, the Moon-forming episode could have turned out very different. Had the impact been better aimed, head-on and dead center, much more of Theia's mass would have wound up as part of Earth. In all likelihood, we would not have a satellite, as Theia and Earth would have merged into one larger moonless world. Or had Theia just missed Earth, its orbit might have been so altered as to be flung inward toward Venus or outward toward Mars, perhaps to leave Earth's neighborhood for-ever. And had the impact been more glancing, the distribution of scattered debris might have produced multiple, albeit much smaller, moons to grace Earth's night sky.

Chance plays a significant role throughout our dynamic cosmic neighborhood. Our Solar System's history is a litany of thwacks and near-misses. The asteroid that helped to kill off the dinosaurs might just as well have been off the mark, saving *Tyrannosaurus* and its descendants to evolve for tens of millions more years. Perhaps big-brained birds would have become intelligent, flying toolmakers. Perhaps the runty mammals of that extended Mesozoic Era would have never amounted to much. With only a little tweak here or there, Earth would have taken a different path.

But some aspects of the cosmos are inevitable, deterministic. The production of huge numbers of protons and electrons, and of corresponding amounts of hydrogen and helium, was hardwired into our universe from the instant of the Big Bang. The formation of stars was an inescapable consequence of the production of huge amounts of hydrogen and helium. The synthesis of all the other elements by nuclear fusion reactions and by supernovas was equally preordained by the formation of hydrogen-rich stars. And the accretion of all sorts of interesting planets—Earth-like, Mars-like, Jupiter-like, and dozens more types that are only now being discovered orbiting distant stars—followed with certainty from the synthesis of all those chemical elements.

So it was that post-Theia Earth settled into a turbulent time of cooling and self-organization. What was that nascent world like? Geologists have dubbed Earth's first five hundred million years the Hadean Eon, in recognition of the hellish conditions that must have prevailed. Informed speculation paints a spectacular picture of Earth's Hadean Eon: sulfurous volcanic exhalations, rivers of glowing lava, and a steady bombardment of asteroids and comets incessantly disrupted Earth's surface. We are nevertheless severely challenged to know Earth's first few hundred million years in any detail, for we are completely lacking in tangible evidence.

Of Earth's origin, we have the rich record of the Solar System—the Sun and the myriad objects bound to it by gravity. Tens of thousands of meteorites provide the most intimate interior glimpses of the age of planetesimals. Details of the Moon's origin are found in every Moon rock and soil. But nothing at all survives from Earth's earliest days, at least nothing known on Earth itself. Not a fragment of rock nor a grain of mineral.

Remarkably, such evidence might still exist in the form of meteorites that were ejected from Earth's earliest surface during giant impacts billions of years ago and subsequently landed back on Earth or on the nearby Moon. Such specimens must exist, perhaps in abundance, some virtually unaltered in all that time. Indeed, the quest for Earth's earliest relics has been cited as one of many scientific rationales for going back to the Moon. Exacting geological surveys of the lunar surface might be lucky enough to find errant Hadean stones and thus reveal truths about Earth's inaccessible past.

But as nice as it would be to hold a piece of Earth's first hardened surface, we aren't completely stymied. For while Earth has changed over and over again, the laws of chemistry and physics have not. Four and a half billion years ago, those laws of chemistry and physics prevailed, as they always do, but without any more really big thwacks or other planet-size complications.

Elemental Inevitability

The early evolution of Earth was a consequence of two intertwined chemical realities: cosmochemistry (making elements) and petrochemistry (making rocks). First came cosmochemistry and the stellar production of all the heavier elements: everything in the periodic table beyond hydrogen and helium, which are elements one and two in the first row. In our universe, several of those chemical elements were destined to become dominant: oxygen, silicon, aluminum, magnesium, calcium, and iron far outweigh all other heavy elements, particularly in the rocky terrestrial planets. Those six elements make up 98 percent of Earth's mass, as they do the mass of Earth's Moon and of Mercury, Venus, and Mars.

Each of these "big six" elements has a distinctive chemical story to tell. Each, in its own way, helped make Earth the way it would inevitably be after the Big Thwack. The key is chemical bonding. Recall that atoms attach to each other when their fuzzy electron clouds interact and shuffle to form more stable arrangements—notably atoms with the magic numbers of two or ten or eighteen electrons. For this kind of exchange to work, some atoms have to give away electrons, while others have to accept them.

Oxygen is Earth's master electron acceptor. Every oxygen atom has eight positively charged protons in its nucleus, which are electrically balanced by eight negatively charged electrons. But oxygen is always looking for two extra electrons to make the magic number of ten electrons. That incessant acquisitive urge makes oxygen one of the most reactive, corrosive gases in nature. It's really rather nasty stuff.

Most of us think of oxygen, first and foremost, as an essential part of the atmosphere (the 21 percent or so that keeps us alive). But that happy atmospheric development is a relatively recent change in Earth history. For Earth's first two billion years at least, the atmosphere was utterly devoid of oxygen. Even today almost all of Earth's oxygen— 99.9999 percent of it—is locked into rocks and minerals. When you hike up a majestic, rugged mountain or walk on a windswept, rocky outcrop, most of the atoms beneath your feet are oxygen. When you lie on a sandy beach, almost two in three of the atoms that support your weight are oxygen.

For oxygen to play this critical chemical role as electron acceptor, there also have to be lots of atoms that can give away or share their electrons. The most prolific electron donor is silicon, which accounts for almost one in every four atoms in Earth's crust and mantle. Silicon has fourteen positively charged protons in its nucleus, which are nominally balanced by fourteen negatively charged electrons. Silicon commonly gives up four electrons to achieve the magic number of ten electrons, becoming a silicon ion with a positive electrical charge. In Earth's rocky crust and mantle, those four relinquished electrons are almost always gobbled up by two oxygen atoms, which become negatively charged ions. Consequently, strong silicon-oxygen bonds are found in almost every rock, most notably in quartz, or SiO_2—a one-to-two mixture of silicon and oxygen atoms. Tough, translucent quartz grains last a long time. They accumulate by the untold trillions along shorelines and are by far the commonest mineral of beach sand. You've also probably seen the beautiful, sharply faceted, transparent crystalline specimens of quartz sold in New Age stores as "power crystals." When you hold such a treasure in your hand, two-thirds of what you're holding is oxygen.

Crystals with silicon-oxygen bonds, collectively called silicates, are Earth's most common minerals, with more than thirteen hundred different known species (and more added almost every month). They are richly varied in their atomic structures and properties because of the versatility of the silicon-oxygen connection, whether in the hearty weather-resistant structure of quartz and feldspar, or the cluster arrangements of gemmy green olivine and red garnet (the semiprecious birthstones of August and January, respectively), or the needlelike chain-silicate habits of some notorious forms of asbestos, or the thin flat sheets of minerals like mica, once used as a cheap substitute for window glass.

Though less abundant than silicon, the elements calcium, magnesium, and aluminum all play key structural roles in the most common silicate rocks throughout Earth's crust and mantle. As positive ions, like their more bountiful silicon cousin, they occasionally bond with oxygen alone, forming the calcium oxide that we recognize as the lawn chemical lime, the rare compound magnesium oxide, and (when trace amounts of the rarer elements chromium or titanium are incorporated into aluminum oxide) the prized gemstones ruby or sapphire.

It is the sixth of the big six elements, iron, that is by far the most versatile. Each of the other five—oxygen, silicon, aluminum, magnesium, and calcium—assumes one dominant chemical personality. Oxygen almost always acts as an acceptor of two electrons, silicon almost always acts as a donor of four electrons, aluminum as a donor of three electrons, and magnesium and calcium as donors of two. But iron, element twenty-six, plays three quite distinct chemical roles.

Iron's versatility is underscored by Earth's layered structure. About one in ten atoms in Earth's oxygen-dominated crust and mantle is iron, whereas Earth's metallic core is more than 90 percent iron. This sharp contrast stems from the fact that this element's twenty-six electrons are a pretty far cry from eighteen, the nearest magic number, making iron a donor par excellence. There's no way iron can give away eight electrons (no one atom will accept that many) so it has to make do with whatever acceptors happen to be present.

Sometimes iron acts just like magnesium and gives up two electrons to become a +2 ion. Iron in this divalent state lends a distinctive greenish or bluish color to many minerals and other chemicals. The characteristic green color of the gemstone peridot (an iron-bearing olivine) and the bluish-green color of oxygen-starved blood in your veins are telltale signs of divalent iron. In this guise, iron bonds to oxygen in a one-to-one ratio. And because magnesium and iron atoms are similar in size, these elements often substitute freely for each other in common minerals of Earth's crust and mantle. Some of Earth's most abundant minerals, including olivine, garnet, pyroxene, and mica, have variants that display pretty much any magnesium-to-iron ratio, from colorless versions with 100 percent magnesium to dark-hued varieties with 100 percent divalent iron.

Iron is not restricted to the +2 state, however. In the presence of lots of electron acceptors, it readily gives up a third electron to become a +3 ion. This trivalent form of iron lends a characteristic brick-red color to its host. Red rust, red soils, red bricks, and oxygen-rich red blood owe their vivid hues to trivalent iron. Like aluminum, which also adopts the +3 state, trivalent iron bonds in a two-to-three ratio with oxygen to form Fe_2O_3—a common mineral named hematite, for its bloodred color. Just as magnesium often proxies for iron's divalent form, aluminum commonly replaces iron's trivalent variant. The minerals garnet, amphibole, and mica display every imaginable aluminum-to-iron ratio, with the iron-rich varieties presenting as red rather than green.

So with the extremely useful trick of switching back and forth from the +2 to the +3 state (we'll come back to this remarkable ability in a couple billion years, when life first comes on the scene), iron in its divalent and trivalent guises acts like the other members of the big six. But wait—iron has one more critical role to play in Earth: it can rather easily form a metal.

Most of the types of chemical bonds introduced so far involve an exchange of electrons, resulting in ions. Aluminum, magnesium, calcium, and iron giveth electrons; oxygen taketh them away. Consequently, these linkages are called ionic bonds. However, metals adopt a very different bonding strategy. In a metal, each atom gives up one or more electrons, to become positively charged. But those disenfranchised electrons hang around in the metal in a kind of sticky, negatively charged sea, which holds all the positively charged atoms together like regimented arrays of little BBs in molasses. Iron metal is a vast collection of iron atoms that collectively share such delocalized electrons.

The consequences of this communal behavior are profound. For one thing, all those shared electrons are free to move around, so metals make excellent conductors of electricity (electricity being nothing more than the controlled flow of electrons). By contrast, in ionically bonded materials made of oxygen plus magnesium or aluminum, every electron is locked into place so tightly that electricity can't possibly flow. Another consequence of metallic bonding is that such materials tend to bend rather than break. The electron sea surrounding the atoms can be folded and twisted without losing its collective strength, quite unlike the behavior of most brittle rocks and minerals.

The perceptive reader will have noted that iron is not alone in performing this metal-forming trick. Aluminum metal cans, foil, and household wiring are commonplace; magnesium metal alloys are a mainstay of high-tech racing cars and other toys; and the semimetal silicon lies at the heart of every electronic gadget (hence Silicon Valley). But metallic aluminum, magnesium, and silicon are modern marvels of the chemical industry. It takes huge chunks of energy to rip those stubborn elements away from oxygen, and their metallic states almost never form in nature.

Iron is less committed to oxygen and more fickle in its chemical bonding partners. Unlike silicon, aluminum, magnesium, or calcium, it is perfectly happy to link to other electron acceptors, notably sulfur—iron sulfide is the shiny mineral pyrite, or fool's gold. Unlike those other elements, iron readily forms a dense metal that sinks to the center of planets and forms their massive cores.

Molten Earth

The big six elements, each of which is an inescapable consequence of the evolution of exploding stars and terrestrial planets, are also responsible for Earth's most abundant rocks. Their distinctive chemical behaviors set our planet on an irreversible course of transformation into the world we inhabit today. But before rocks could form, Earth had to cool.

Imagine once again the violent years following the Moon-forming impact. For a few days or weeks, what would become Earth and what would become the Moon were still being sorted out. Neither Earth nor the Moon in those early post-Theia days had any solid surface. These coalescing companion globes were both bounded by an encircling magma ocean, roiling and incandescent red, pelted by an incandescent molten silicate rain at temperatures of thousands of degrees.

As the air cleared of Theia's remains, blast-furnace-like heat radiated from Earth into the cold vacuum of space, inexorably cooling the planet's outer shell. Even so, cosmic events conspired to keep Earth's surface molten a while longer. For one thing, big asteroids kept pounding the planet. Each collision added more thermal energy, superheating the impact area, thwarting any attempts to form a stable crust. Intense gravity-induced tides from the nearby Moon also helped to maintain Earth's liquid state, as an equatorial bulge of turbulent magma swept around the planet every five hours, fragmenting the organization of any thin, solid veneer. Earth's ample store of highly radioactive elements—both the short-lived, heat-generating isotopes of aluminum and tungsten and the long-lived radioactive isotopes of uranium, thorium, and potassium—continued to add even more heat. And a young and growing atmosphere, fueled by the volcanic release of vapors rich in carbon dioxide and water, may have amplified these effects by inducing a "super greenhouse" state.

For an unknown length of time, perhaps a hundred years or a hundred thousand years—a geological blink—the molten state prevailed. But cooling and hardening were preordained. The second law of thermodynamics demands that hot objects, lacking significant new energy inputs, must cool, and the hotter the object, the faster the rate of cooling.

Three familiar mechanisms facilitate this transfer of heat. First there's conduction. When a hotter object touches a cooler object, heat energy must flow from hot to cool. This process, painfully obvious if your feet have ever been burned from walking on sunbaked pavement or your hand has ever blistered from touching a stove burner, results from the constant twitching of atoms. Atoms in hotter objects experience more violent motions. When a cooler object, with more slowly wiggling atoms, contacts a hotter object, with more frenetic atoms, some of that violent motion is transferred by atom-to-atom collisions. If the hot object you touch is hot enough, it can disrupt the molecules in your skin, killing cells, causing a burn. Conduction is a fine way to transfer heat locally, from one object to an adjacent object, but it is a poor choice for heat transfer on a planetary scale. It just takes too long to move heat from one wiggling atom to the next.

Convection, by which collections of hot atoms move thermal energy wholesale, is a better planetary choice for cooling. You experience convection whenever you boil water. Pour water in a pan, turn up the heat, and wait. The process is slow at first, as the hot pan contacts the cold water and transfers heat by conduction—wiggle by wiggle, metal atoms in the pan jostle molecules of water. But soon another mechanism takes over. Heated volumes of water on the bottom begin to expand and rise through the cooler, denser overlying water, transferring heat en masse to the surface. Simultaneously, the cooler, denser surface waters sink to the hot bottom. Faster and faster the heat exchange progresses, with columns of water rising and sinking, until you reach a rolling boil. Through the convective cycling of hot water up and cool water down, large volumes of water spread heat throughout the liquid in a fast and efficient dance.

On the grand scale of Earth, convection appears over and over—in cooling offshore breezes on a summer day, in grand ocean currents that sweep from the Equator to the Arctic, in the turbulent lightning-laced fronts of thunderstorms, in boiling hot springs and spurting geysers. And so it is in Earth's interior, where hot pressurized rocks soften and flow like taffy over spans of millions of years. Cooler, denser rocks near the surface sink, as hotter, less dense rocks rise to take their place. For all of Earth's history, convection has been the primary driver of planetary cooling.

And then there's radiation, the third mechanism of heat transfer. Any hot object radiates heat to its cooler surroundings in the form of infrared radiation that travels 186,000 miles per second through a vacuum. This familiar form of energy, abundantly evident whenever you relax and soak up the rays of the shining Sun, is similar in its behavior to waves of visible light (though heat radiation has slightly longer wavelengths). Perhaps the most obvious infrared energy source is the Sun, which bathes Earth in infrared radiation that travels across the vacuum of space in about 8.3 minutes. An electric space heater, a toasty fire in your fireplace, and an old hot water radiator are other familiar examples. Every warmer object radiates heat to its cooler surroundings. Your body is no exception. That's why a crowded auditorium can get so uncomfortably warm—each person radiates heat like a hundred-watt lightbulb—a fact one may easily verify by putting on a pair of night-vision goggles, which make people and other animals that emit infrared radiation appear to glow brightly in the dark.

The rate of heat transfer, whether by conduction, convection, or radiation, depends on the difference in temperature between the hotter and colder objects. Conduction is swifter, convection more vigorous, and radiation much more intense if the temperature differences are large. Earth is a warm planet. Orbiting the Sun as it does in the coldness of space, it is always radiating heat into the void. But the red-hot post-Theia Earth blasted its excess heat energy into space at a rate unmatched in modern times. It literally glowed in the black void of space.

First Rocks

Given Earth's prodigious heat loss to space, the formation of a rocky crust was inevitable. Somewhere, probably near one of Earth's less tidally stressed poles, the molten surface cooled just enough for the first crystals to form. But cooling and crystallizing was far from a simple event. Many everyday substances have a well-defined temperature at which a cooling liquid becomes solid—the familiar freezing point. Liquid water freezes at 32 degrees, silvery mercury metal at −38 degrees, and ethanol (the common alcohol in booze) at −179 degrees Fahrenheit. But magma is different. It is a curiosity of magma that it doesn't have one single freezing point (although *freezing point* in the context of magma at more than 2,500 degrees Fahrenheit seems something of an oxymoron).

Let's begin with the immediate post-Theia inferno 4.5 billion years ago, a time when Earth and its Moon shared a radiant silicate vapor atmosphere at temperatures of 10,000 degrees Fahrenheit. That hellish rock gas cooled rapidly and eventually condensed into droplets and rained magma onto the new twin worlds, as it inexorably cooled to below 5,000 degrees, then 4,000 degrees, then 3,000 degrees. That's when the first crystals started to form.

Such stories of Earth's first rocks are the scientific domain of the experimental petrologists, women and men who devise novel lab techniques to bake and squeeze rocks in order to mimic conditions of Earth's deep interior. The quest to discover rock origins faces two technical challenges. First, you need to control incredibly high temperatures of thousands of degrees, far hotter than any oven or furnace in your home. To do so, scientists craft platinum wire into meticulously spaced coils, through which they deliver high electrical currents to achieve temperature extremes. Even more challenging, these temperatures must be applied while samples are subjected to crushing pressures tens or hundreds of thousands of times that of the atmosphere. For this exacting task, researchers enlist massive hydraulic rams and giant viselike presses.

For more than a century, the Carnegie Institution's Geophysical Laboratory, my scientific home, has been a center for these heroic quests for Earth's deep truths. For a short while, before his untimely death by hospital, I had the chance to work side by side with Hatten S. Yoder, Jr., one of the pioneers of experimental petrology and the world's foremost expert on the origins of basalt. Imposing, dynamic, enthusiastic, and attentive, Yoder was literally a towering figure in the field. As a naval officer in World War II, he was intimately familiar with gigantic metal hardware. In the 1950s, he joined the Geophysical Laboratory and used naval surplus gun barrels and armor plating, still painted battleship gray, to build the high-pressure lab that would frame not just his half-century career but our comprehension of the ground we stand on.

The centerpiece of Yoder's device was a "bomb"—a massive steel cylinder a foot in diameter, twenty inches long, with an inch-diameter bore. One end of the bomb was connected to a series of gas pumps, compressors, and intensifiers that could generate a staggering twelve thousand atmospheres of gas pressure—the pressure found twenty-five miles beneath Earth's surface—with a pent-up energy equivalent to the explosive power of a stick of dynamite if the apparatus ever failed catastrophically. The bomb's other open end accepted a foot-long rock sample assembly and a giant hexagonal nut, six inches in diameter. We sealed the device by tightening the nut with a three-foot-long, twenty-pound wrench.

The beauty of Hat Yoder's apparatus was that we could load powdered rock and mineral samples into little gold tubes, pack the tubes into a cylindrical heater, and secure the entire assembly inside the bomb's pressure chamber. Pump up the pressure, turn on the electric heater, and the bomb did all the work. Each experimental run held up to six small gold tubes; each run lasted from a few minutes to a few days. Hat Yoder's remarkable invention was ideally suited to study how rocks evolve in Earth's crust and upper mantle.

What Hat Yoder and his colleagues found was that an incandescent melt rich in the big six elements will typically begin to solidify by forming crystals of the magnesium silicate olivine as it cools below about 2,700 degrees Fahrenheit. On both Earth and the Moon, during that long-ago cooling period, beautiful tiny green crystals began to grow in the magma as microscopic seeds, which expanded to the size of BBs, peas, grapes. But olivine is typically denser than the liquid in which it grows, so those first crystals began to sink, faster and faster as the crystals grew larger and larger, accumulating huge deep masses of nearly pure crystals—forming a stunning green rock called dunite. On Earth this rock is relatively rare today, appearing at the surface only on the special occasions when mountain-building activities of uplift and erosion expose the distinctive dense, deep-formed olivine cumulates.

The continuous sinking of olivine crystals gradually altered cooling magmas inside Earth and the Moon. The remaining hot melts changed composition; as they became progressively depleted in magnesium, they became correspondingly more concentrated in calcium and aluminum. On the Moon, as the magma ocean continued to cool, a second mineral started to form. Anorthite, a feldspar made of calcium aluminum silicate, began to crystallize alongside olivine, forming pale blocks. Unlike olivine, anorthite is less dense than its host liquid, so it tends to float. On the Moon, immense quantities of anorthite popped to the surface of the magma ocean to form a vast crust of floating feldspar mountain ranges rising as much as four miles above the molten surface. These whitish-gray masses, which still dominate 65 percent of the Moon's reflective silvery face, are called the Lunar Highlands. Rising as they did directly from the magma ocean, they are the oldest known formations on the Moon. Apollo samples reveal a range of ancient ages for these distinctive anorthosites from as young as 3.9 billion years to almost 4.5 billion years, shortly after the Big Thwack.

On Earth, with its wetter composition, deeper magma oceans, and correspondingly greater internal temperatures and pressures, a somewhat different scenario unfolded. A small amount of anorthite probably crystallized early in Earth history in some near-surface, low-pressure environments, but it was a rather minor mineral. Instead, magnesium-rich pyroxene, the commonest of the chain silicate minerals, appeared in abundance, to commingle with olivine in a thick crystal slush. Earth's earliest rocks thus predominantly featured olivine and pyroxene in a hard, greenish black rock called peridotite. Varieties of peridotite began to crystallize throughout Earth's outer fifty miles, probably commencing more than 4.5 billion years ago and continuing for many hundreds of millions of years.

In spite of its early abundance, peridotite, too, is relatively rare at Earth's surface today. By one persuasive scenario, rafts of peridotite hardened and cooled to form Earth's first transient rigid surface. But cooling peridotite, like its dunite predecessor, is significantly denser than the hot magma ocean in which it formed. The peridotite surface layer thus cracked, buckled, and sank back into the mantle, to displace more magma that cooled to form more peridotite. Over a span of hundreds of millions of years, the mantle itself slowly solidified, riding a kind of peridotite conveyor belt that operated in Earth's outer fifty miles. The ratio of dense solid peridotite to magma increased, until the upper mantle was mostly solid olivine-pyroxene rock.

Core Truths

Deeper in the mantle, fifty to two hundred miles beneath the crust, cooling and crystallization must have proceeded in a similar fashion, albeit more slowly. Details of the process remain uncertain—the next generation of high-pressure, high-temperature apparatus must be brought to bear to sort out the complexities—but separation of crystals from melts by sinking and floating probably played a significant role, as they did in the nearer surface environment.

Much of what we know of those hidden, deep domains comes from the science of seismology, the study of sound waves speeding through Earth's deep interior. Earth is constantly ringing like a bell: crashing tides, rumbling trucks, and earthquakes both big and small all conspire to shake Earth and propagate seismic waves. And like sound waves in a steep-walled canyon, seismic waves echo when they bump into a surface. Seismic waves reveal that Earth's interior is a complexly layered place.

At the most basic anatomical level, Earth is triply layered—it has a thin, lower-density crust at the surface, a thick, higher-density mantle in the middle, and a thicker, really dense metallic core in the center. Each of those three domains contains further layering. The mantle, for example, is divided into three sublayers—upper mantle, transition zone, and lower mantle. The peridotite-dominated upper mantle extends down perhaps 250 miles, at which depth pressure forces the atoms in olivine to pack into a denser silicate crystal form called wadsleyite, the dominant mineral of the mantle's transition zone. The lower mantle, 150 miles farther down, features an even denser assemblage of magnesium silicates. The pressures in the lower mantle are so high—hundreds of thousands of times the surface pressure—that silicon-oxygen bonds adopt an even denser, more efficient packing arrangement of atoms called perovskite.

Seismic studies document the nature and extent of each of these mineralogically distinct mantle layers, and by and large the transitions from one to the next are neat and tidy. The exact depths of the transitions vary slightly by ten or twenty miles from place to place—beneath the continents versus the oceans, for example—but each boundary appears to be relatively smooth and well behaved. By contrast, seismology provides tantalizing evidence to suggest that the core-mantle boundary is an especially complicated zone, rather different from the clean mantle-mantle transitions. To a first approximation, the core-mantle boundary produces the expected strong echo. Indeed, the density contrast between silicate mantle and metal core is so extreme as to create a physical boundary as sharp as that between air and water—producing the strongest reflected seismic signal from Earth's deep interior. More than a century ago, that divide was one of the first hidden features of Earth's deep interior that seismologists discovered.

A perfectly smooth and regular boundary would produce a sharp, focused seismic reflection—an echoing response that could be recorded as a distinctive spike on a seismometer. But seismic signals reflecting off the core-mantle boundary are often messy, smeared out, broken up. There's extra structure down there, like irregular lumps or piles of debris. Geophysicists, who are not always known for employing the catchiest terminology, call this lumpy chaotic zone the D″ (D double prime) layer. (Astrophysicists, who coined such imaginative terms as *brown dwarf, red giant, dark energy,* and *black hole,* are rather more successful at the scientific name game.)

The complexity of this deep D″ feature is in part the result of the sharp density contrast between the core's homogeneous iron metal and the mantle's varied oxygen-rich minerals. All mantle minerals float on the dense core like corks on water, but these diverse minerals can differ widely in their densities. In the primordial magma ocean, some silicates sank, and some floated. As a result, big chunks of the earliest crystallized solids sank through the mantle all the way down, to float like rafts on the metal core. Some seismologists envision three-hundred-mile-high "mountains" with irregular piles of dense minerals resting on the core-mantle boundary, where they chaotically deflect seismic signals.

Surprisingly, there may also be big core-mantle boundary puddles and ponds of unusually dense silicate liquid, perhaps rich in the elements aluminum and calcium as well as a slew of "incompatible elements" that seem to be missing from inventories of Earth's outer layers. We have no easy way to be sure, but seismologists point to deep, localized "ultra-low velocity zones," in the D″ layer just above the core-mantle boundary, where seismic waves travel about 10 percent slower than they do in adjacent rocks. Slow seismic waves are often a telltale sign of liquid. Such deep liquid lakes and ponds could also provide a neat solution to that niggling missing-element problem: just stick all the incompatible elements in the inaccessible D″ layer, where they are forever sequestered in that enigmatic, heterogeneous zone of mineralogical junk.

And what of the core itself? When Earth was very young, the dense, iron-rich core, more than 2,000 miles in diameter, had fully formed and was probably entirely molten (unlike today, when the inner core appears to be a growing ball of solid iron crystals 750 miles in diameter). Temperatures at that sharp core-mantle dividing line may have exceeded 10,000 degrees Fahrenheit, while pressures exceeded a million times that of our modern atmosphere.

The hot core was from the outset, and remains to this day, a dynamic place of swirling liquid metal currents. One important consequence of these currents was the early generation of Earth's magnetic field—the magnetosphere, which is like an immense electromagnet. Magnetic fields bend the paths of electrically charged particles, so Earth's magnetosphere provides a kind of invisible deflector shield to the intense bombardment of solar wind and cosmic rays—a barrier that is perhaps a prerequisite for the origins and survival of life.

The core is also an important source of heat energy that helps to drive convection in the mantle. Even today plumes of soft, plastic rocks from the core-mantle boundary rise almost two thousand miles to the surface in volcanic hot spots such as Hawaii and Yellowstone. Remarkably, the fixed locations of these plumes on the surface may be dictated by deep topography. The three-hundred-mile-tall mountains of the D″ layer may act as thermal blankets lying on the hot core, so it's possible that hot spots originate at the deepest, heat-releasing valleys between those epic, hidden mountains.

Basalt

At heart, the mineral evolution story rests on a preordained succession of rock types, each mineral-forming stage following logically from the previous stage. Earth's first peridotite crust was a critical but fleeting juvenile phase, born of the primordial magma sea. When it ultimately cooled and hardened, it proved too dense to remain anywhere near the surface and thus sank back into Earth's depths. Another, less dense rock was required to girdle the globe. That rock was basalt.

Black basalt dominates the near-surface rocks of every terrestrial planet. The asteroid-scarred exterior of Mercury is mostly basalt. So are the scorched, mountainous surface of Venus and the weathered red surface of Mars. The Moon's darkly splotched mares (seas), which contrast so vividly with the paler gray anorthosite highlands, are the hardened remains of immense black basaltic lakes. And on Earth, 70 percent of the surface, including all the floors of all the oceans, is underlain by basalt crust.

Basalt comes in a variety of flavors, but two essential silicate minerals dominate them all. One key mineral is plagioclase feldspar, by far the most important aluminum-bearing mineral in terrestrial planets and moons and Earth's commonest crustal mineral. My MIT professor Dave Wones once advised me that if I was ever shown a mystery rock and quizzed as to its mineralogy, I should just say "plagioclase," and I'd be right 90 percent of the time. The second essential mineralogical ingredient of basalt is pyroxene, the common chain silicate also found in peridotite. Pyroxene is one of a handful of common minerals that can incorporate all of the big six elements (and many more less common elements, as well).

To understand the origins of plagioclase and pyroxene, the two essential mineral ingredients of basalt, consider the strange freezing and melting habits of rocks. Four and a half billion years ago, as Earth's magma ocean cooled, olivine formed first, then a little bit of anorthite, and finally a lot of pyroxene. The resulting magnesium silicate rock was peridotite, which formed much of the upper mantle. As masses of peridotite formed and sank, they were reheated and partially remelted.

Our everyday experience with melting suggests that the change from solid to liquid takes place at one specific temperature. Water ice melts at 32 degrees, most household candle wax at about 130 degrees, and dense metal lead at 621 degrees Fahrenheit. But rock melting isn't so simple; most rocks don't melt entirely at one temperature. If you heat peridotite to about 2,000 degrees Fahrenheit, the first melt will appear. (It will appear sooner if the peridotite is rich in volatile water and carbon dioxide.) The composition of those first microscopic droplets differs dramatically from that of the bulk of peridotite rock. The initial melt has a lot more calcium and aluminum, a little more iron and silicon, and a lot less magnesium. This initial liquid is also a lot less dense than its peridotite host. Consequently, even a 5 percent melting of peridotite in the mantle generates a lot of magma that gathers along mineral grain boundaries, collects in fissures and pockets, and rises toward the surface—magma that will eventually become basalt. Over billions of years of Earth history, the partial melting of peridotite has generated hundreds of millions of cubic miles of basalt magma.

Molten basalt comes to planetary surfaces in two complementary ways. The more spectacular is through volcanic eruptions like those in Hawaii and Iceland, with fiery magma fountains and riverlike flows. Such dramatic eruptions are a consequence of water and other volatiles, which remain locked in the silicate liquid at the high pressures more than a mile down, but which transform explosively to gas near the surface. Such explosive volcanism can eject ash and toxic gases high into the stratosphere and can hurl car-size volcanic "bombs" more than a mile outward to smush the surrounding countryside.

Layer by layer, these basalt eruptions of lava and ash can build mountains several miles tall and cover thousands of square miles in black rock. This type of basaltic lava flow and ash is extremely fine-grained and rich in glass, the consequence of liquid cooling so rapidly that crystals don't have time to form. The result is a featureless black crust of hardened lava. Other distinctive olivine basalts, which occur only when peridotite is partially melted at relatively shallow depths of less than twenty miles, contain a few lustrous olivine crystals that formed underground in the first stage of melt solidification. The green crystals decorate the otherwise bland black rock.

It takes a lot of explosive force for magma to break through to the surface, so a significant fraction of basaltic magma never makes it aboveground. Rather, these red-hot liquids are stuck far underground, where they cool more slowly and form inch-long, lathe-shaped feldspar and pyroxene crystals in rocks called diabase or gabbro. Sometimes the magma is injected into near-vertical cracks in subsurface rocks to form smooth-faced dikes. If the host rock is soft and erodes away millions of years later, the result can be a long, straight diabase wall that can look uncannily like a crumbling archaeological site. Alternatively, if the magma is injected between flat-lying layers of sedimentary rocks, it can form a thick blanketlike sill. The Palisades cliffs on the Hudson River, prominent just upstream from New York City on the western shore of the Hudson River, are the result of one of a series of basaltic sills that dip gently to the west to form parallel highlands (and some of the most expensive real estate) in northern New Jersey and southern New York. Still other times the liquid sits and cools in irregular magma chambers that can form miles underground and that stretch for miles across. But whatever their ultimate geometry, diabase and gabbro are really exactly like basalt.

With the inevitable formation of the basaltic crust, Earth for the first time enjoyed a sturdy, solid surface that could float. Before the crust, when magma and peridotite alone defined the planet's surface, no topographic feature could rise for long to a significant height above the mean elevation. Red-hot peridotite mush is not nearly strong enough to support a mountain. But tough basalt, with its relatively low density, is a different story. The average density of basalt is more than 10 percent lower than that of peridotite. That means a floating mass of basalt ten miles thick will project more than a mile above the magma ocean. Rapidly accumulating volcanic cones could rise even higher, perhaps more than two miles above the mean. As a consequence, Earth's blemished surface began to develop some real character.

Hostile World

Seen from space—from the safe distance of the young Moon, for example—Earth's basaltic veneer appeared deep black with arcuate red cracks and local bright points where immense fountaining volcanoes broke the surface. Jets of dirty white, ash-laden steam obscured some of the most volatile-rich volcanic cones and adjacent portions of the globe.

Imagine yourself back more than 4.4 billion years in time to the newly minted black surface of that Hadean Earth. You could not have survived long in the harsh, alien landscape. Meteors incessantly bombard the surface, cracking the thin brittle black crust, showering shattered rock and gobs of magma across the plains. Countless volcanic cones rise, growing steadily to heights of many thousands of feet, their immense magma fountains powered by the explosive release of steam and other volatiles that will, some fine day, cool enough to become the oceans and atmosphere. No trace of life-supporting oxygen is to be found. On this unforgiving young Earth, your nostrils are assaulted by foul-smelling sulfurous compounds, your skin scalded by venting steam, your eyes burned by the noxious hot gases. Your excruciating death agonies will be brief on such a hostile world.

The receding Moon continued to play a major role in shaping the crust. Globe-spanning tides of rock and magma, though less extreme than in the first centuries following Theia's demise, repeatedly cracked and buckled the surface, opening fissures that oozed red-hot rocky mush and thwarted the formation of a solid surface. The Moon's uncomfortable proximity also perpetuated Earth's insanely rapid rotation—five-hour days persisted, accompanied by megastorms and ultratornadoes far more severe than anything hyped on today's Weather Channel.

But beneath that wretched surface, Earth's inexorable evolution to a living world had begun. The well-mixed, molten interior began to separate into volumes of distinctive compositions—matter that would become the continents and deep-sea crust, the atmospheres and oceans, plants and animals. Heating and cooling and crystallization, crystal separation by settling and floating, accumulation of peridotite, partial melting—these processes shaped Earth through its infancy 4.5 billion years ago, and they persist even unto the present day.

Earth's vast reservoir of internal heat, the central theme of this chapter, continues to play the dominant, transformative role in shaping our planetary home. Today the most obvious manifestations of this deep, hot realm are intermittent volcanoes, with their fiery magma fountains and red-hot rivers of molten rock. Erupting geysers and sulfurous hot springs also hint at a hellish hidden subsurface realm. Throughout Earth's 4.567-billion-year history, as heat inexorably worked its way outward from the incandescent center to the fractured crust and thence into the coldness of space, the surface has borne the brunt. Buffeted by the swirling convection of the mantle and stressed by the Moon's incessant pull, the crust has bent and buckled, cracked and twisted. Continents have constantly shuttled across the globe, ripping apart, colliding, and scraping past one another in the ongoing heat-driven dance of the tectonic plates. Every day of our lives, Earth's inner heat reworks the rocks on which we live, recycles the water we drink, and alters the air we breathe.

Because of heat, Earth was destined for a brief time to be a black world, glazed with a thin basaltic veneer. But that brief juvenile phase could not last long. A new volcano-born layer of brilliant blue was about to gird the globe.

EARTH'S AGE *(billions of years)*

0	1	2	3	4	4.567
Hadean Eon	Archean Eon	Proterozoic Eon		Phanerozoic Eon

↑

Chapter 4

Blue Earth

The Formation of the Oceans

Earth's Age: 100 to 200 million years

❧ Earth's infancy, its first half-billion years or so, is shrouded in mystery. Rocks and minerals provide tangible evidence for most of our planet's storied past, but few rocks or minerals survive from that most ancient Hadean time. As a consequence, any narrative of Earth's initial cooling and the subsequent watering of its black surface must be based on speculations informed by experiments, models, and calculations. Even so, some uncertainties will always remain.

That's not a bad thing. What makes each day at the lab new and exciting is the richness of "what we know we don't know" and the possibility each day that we will discover some small clue that brings us closer to truth. Even more tantalizing is the prospect of discovering aspects of the natural world that "we didn't know we didn't know"— discoveries that increase the breadth of mystery.[*] These new ways of

---

[*] Though often attributed to Donald Rumsfeld's 2002 speeches, those quotes first appeared several years earlier in the preface to my 1997 book with Maxine Singer, *Why Aren't Black Holes Black?*

asking questions—"How did minerals evolve?" for example, rather than simply "What are their chemical and physical properties?"— pave the way for breakthroughs.

It's important to take an inventory of what we don't know. All the evidence suggests that the Moon formed by an epic impact, yet we can't be sure exactly when the collision occurred, nor what were the nuances of Theia's final trajectory. Following that colossal collision, we can imagine an incandescent torrential rain of silicates onto Earth's tortured magma ocean, but the duration and rate of cooling of such a superheated world are poorly constrained and will remain topics of much debate for decades to come. The proximity and recession rate of the newly formed Moon, though critical to understanding the dynamics and evolution of early Earth, are equally uncertain. Likewise, no one knows when the oceans first formed, nor exactly what they looked like. But form they did, and the following story is based on the best evidence available and so is as good as it gets for the time being.

The black Earth could not remain black for long. Global-scale volcanism spewed hot nitrogen, carbon dioxide, noxious sulfur compounds, and water vapor into the thickening atmosphere at rates of billions of tons per day. Those volatile elements and compounds—the very same molecules that formed the varied ices of the former nebula, the very same atoms that you are now breathing and that make up the intricate tissues of your body—played many roles in the rapidly evolving Earth. When hot water mixed with rock magmas, it lowered their melting temperatures and turned them into a superheated soup that rose toward the surface. Close to the surface, the gases dissolved in that magmatic soup transformed from liquid to violently expanding gas in massive volcanic explosions, much as a shaken soda will blast its way out of a confining can. Water-rich fluids also dissolved and concentrated rare elements—beryllium, zirconium, silver, chlorine, boron, uranium, lithium, selenium, gold, and many more—that would eventually become the great ore bodies of Earth's diversifying crust. At the chaotic surface, roaring rivers and crashing waves became principal agents for the erosion of rock, the formation of Earth's first sandy beaches, and the accumulation of thickening near-shore wedges of sediments. In short, water became the chief architect of Earth's solid surface.

Any focus on oceans and atmosphere reflects a somewhat anthropocentric view, for these fluid bodies are trivial components of the planet as a whole. Oceans today represent only about 0.02 percent of Earth's total mass, while the atmosphere is no more than one part per million of its bulk. Nevertheless, oceans and atmosphere have exerted, and continue to exert, disproportionately large influences in making Earth the unique world that it is.

Five principal players—nitrogen, carbon, sulfur, hydrogen, and oxygen—take the leading roles as Earth's mobile, gaseous components. All of these ingredients are produced abundantly in large stars, all are widely dispersed in supernova explosions, and all became concentrated in the most primitive carbon-rich chondrite meteorites more than 4.56 billion years ago.

In many respects, the average composition of chondrite meteorites closely matches that of Earth today. The big six elements discussed in chapter 3 (oxygen, silicon, aluminum, magnesium, calcium, and iron) are remarkably similar in their proportions, as are numerous less common elements. But even a cursory study of those fascinating ancient objects reveals that much of Earth's original inventory of volatiles is missing from today's planet. The most primitive chondrites average more than 3 percent carbon, but all the known carbon reservoirs on Earth now add up to less than 0.1 percent. Similarly, the water content of chondrites is much greater than Earth's modern average—perhaps a hundred times more. Such gross compositional disparities point to a chaotic and violent past. Most of Earth's volatiles must have been lost to space or deeply buried, far beyond our ability to sample.

The key to understanding Earth's early transformation from a blasted, inhospitable black planet to a cooler, habitable blue planet lies in the story of its peripatetic volatiles. But no volatiles survived in an unaltered state from the Earth's first half-billion years. Almost all the nitrogen and carbon, all the sulfur and water, have been recycled countless thousands of times, as the same atoms have been used over and over again. Chondrite meteorites provide a quantitative starting point for our guesswork; the few known rock and mineral samples from Earth's first billion years, coupled with data from the Moon and other objects in the Solar System, further delineate our speculations. And as was the case with understanding the evolution of mantle and crust during Earth's first hundred million years, as well as the formation of stars long before that, the key to any viable scenario is knowledge of the immutable characteristics of the elements in question—in this instance, the physical and chemical properties of volatile nitrogen, carbon, sulfur, and water.

Of these four ingredients, nitrogen is the easiest to deal with. It's a chemically inert gas that forms few minerals, plays almost no role in rock formation, and tends to concentrate in the atmosphere. Only since the rise of life has the nitrogen cycle had much effect on Earth's outer layers. Carbon and sulfur would also come into much greater prominence one or two billion years into Earth's existence, when life and an oxygen-rich atmosphere transformed the surface realm. But the fourth ingredient, water, has been central to Earth's story from the get-go.

Water: A Short CV

Water's varied geological roles follow from the unique chemical properties of hydrogen oxide. Recall that hydrogen is element number one, while oxygen is element eight; neither element has a magic number of two or ten electrons. Each electron-accepting oxygen atom seeks two more electrons to reach the magic number ten, while each hydrogen atom with one electron to share wants one more. The molecular result is a two-to-one ratio of hydrogen to oxygen: H_2O. The atoms in this compact unit form a V shape: the central larger oxygen atom is flanked by two hydrogen bumps, not unlike Mickey Mouse's ears. The oxygen atom, having borrowed electrons from two hydrogen atoms, assumes a slight negative electrical charge, while each hydrogen atom is correspondingly slightly positive. The result is a polar molecule, with opposed positive and negative electrically charged parts (Mickey's ears and chin, respectively).

Such polarity in water's molecules accounts for many of its distinctive properties. Polar water is a super solvent, because its positive and negative ends exert strong forces that can pull apart other molecules. Consequently, table salt, sugar, and many other ingredients dissolve rapidly in water. Most rocks take a bit longer to dissolve, but over millions of years, the oceans have become rich in almost all the chemical elements. (As a consequence, every cubic mile of ocean water holds about four hundred pounds of gold—more than $10 million worth at the precious metal's recent high value, if only we had the technology to extract it.) This unparalleled ability of water to dissolve and transport other chemicals also makes it an ideal medium for the origins and evolution of life. All life on Earth, and perhaps all life in the cosmos, depends on water.

The polarity of water molecules causes them to bond strongly to one another: the positive side of one molecule attracts the negative sides of other molecules. Consequently, ice is an unusually strong molecular solid (a fact you'll not soon forget if you've ever fallen hard while ice-skating). Unusually strong intermolecular bonding also results in water's unusually high surface tension—a fascinating property that allows small insects literally to walk on water. Surface tension also leads to capillary action, which causes water to rise through the stems of vascular plants and allows trees to soar hundreds of feet above the land. Rounded raindrops, pulled together by the strong mutual attraction of water molecules, are yet another manifestation of surface tension and a vital link in maintaining Earth's unusually rapid water cycle. Nonpolar volatile molecules like methane and carbon dioxide can't form such droplets. They would just float in the atmosphere as an ultrafine, pervasive mist, so "rain" would be unknown on a planet dominated by those atmospheric gases.

Strong bonds between molecules result in another of water's most curious and important properties: liquid water is about 10 percent denser than solid ice. In almost every known chemical compound, the solid sinks in its liquid—a situation that is intuitively logical because the regular, repeated packing of molecules in solids contrasts with their random distribution in liquids. Think about storing shoe boxes in the back room of a shoe store. Neat stacks and rows of boxes (like perfectly aligned molecules in a solid crystal structure) take up much less volume than a random pile (like chaotically tumbling molecules in a liquid). But in water, the molecules actually pack more efficiently in their random liquid state than they do in orderly ice crystals.

The important consequence is that ice, whether it is a cube in your drink, a layer on a frozen river or stream, or a giant iceberg, floats. Were it not for this unusual characteristic, many bodies of water would freeze solid, bottom to top, rather than forming a thick, protective surface layer of ice every winter. In such a solidly frozen world, aquatic life in cold ecosystems might be severely challenged, while the vital water cycle would all but come to a halt. Curiously, the same phenomenon is one of several factors (albeit a minor one) that facilitate ice-skating and skiing. The high pressure exerted by the blade of your skate pressing down on solid ice helps to produce a thin layer of denser liquid water over which you can glide. If the temperature becomes too cold, typically below about −100 degrees Fahrenheit, the lubricating water layer doesn't form, and ice-skating and skiing become much more difficult.

Yet another distinctive characteristic of "pure" water is its lack of purity. No matter how carefully filtered or distilled, water is *never* made entirely of H_2O molecules. A small fraction of those three-atom units inevitably splits apart into positively charged hydrogen ions (hydrons, or H^+ ions, which are actually just individual positively charged protons without any electrons attached), plus negatively charged hydroxyl groups (OH^- ions). Hydrons quickly latch on to water molecules to produce H_3O^+ hydronium ions. What we call pure water at room temperature contains equal numbers of positive hydronium and negative hydroxyl groups, at a concentration that translates to a pH of 7 (a "power of hydrogen" of 10^{-7} moles of hydronium groups per liter, in chemistry terms).

A prime focus of speculation on Earth's earliest oceans is their pH and salt content. Water easily dissolves lots of impurities, some of which are positively charged like sodium (Na^+) or calcium (Ca^{2+}) ions, and others negatively charged like chlorine (Cl^-) or carbonate (CO_3^{2-}) ions. A rule of thumb is that the net electrical charge of any bulk water solution must be zero: the total number of positive charges must be balanced by an equal number of negative charges. In pure water at room conditions, 10^{-7} moles of H_3O^+ is perfectly balanced by 10^{-7} moles of OH^-. In acids, however, excess H_3O^+ has to balance negative ions (such as chlorine in hydrochloric acid, HCl). In bases, excess OH^- has to balance positive ions (such as sodium in the base sodium hydroxide, NaOH).

The strength of acids and bases is quantified by the pH scale. Smaller values of pH point to acidic solutions with more H_3O^+ than OH^- ions. A slightly acidic solution with pH 6 (typical of untreated tap water in many localities) has ten times more hydronium ions than a neutral solution at pH 7. More strongly acidic liquids include coffee (pH 5, with a hundred times more H_3O^+), vinegar (pH 3, with ten thousand times more H_3O^+), and lemon juice (pH 2, with one hundred thousand times more H_3O^+). By contrast, bases are liquids with more OH^- than H_3O^+, and thus with pH values greater than 7.

Common bases include baking soda (pH 8.5), milk of magnesia (a common antacid with pH 10), and household ammonia cleaners (pH 12). As we shall see, the pH and salinity of Earth's earliest oceans remains a hotly debated matter.

Water, Water Everywhere

Water is one of the most abundant chemicals in the cosmos. The more we look, the more we find, and its presence on other planets and moons and comets provides clues to its abundance on Earth, as well as the possible distribution of water-dependent life in the universe. Telescopic observations can be tricky, since our water-rich atmosphere tends to obscure all but the most concentrated deposits of H_2O on more distant sources. Nevertheless, some objects in deep space reveal an icy surface via their distinctive absorption of infrared radiation.

This spectroscopic fingerprint reveals that some comets and asteroids incorporate significant amounts of frozen water. Astronomers have documented numerous icy worlds in our Solar System, from Pluto and its companion moon Charon, to Saturn's luminous ice rings. All of the gas giant planets, though primarily made of hydrogen and helium, hold significant stores of water vapor in their dense atmospheres. And Jupiter's large moons Europa and Callisto are now thought to feature an ice veneer a few miles thick over much deeper encircling oceans of water.

Closer to home, the other terrestrial planets at first glance appear to be rather dry. Nevertheless, recent observations by NASA's *Messenger* mission to Mercury have identified substantial ice deposits in frigid polar craters that are permanently shielded from the nearby Sun. Venus, the next planet out, may initially have had an Earth-like share of water, but today it appears to be almost totally lacking in near-surface H_2O. Its thick, superheated carbon dioxide atmosphere speaks of a runaway greenhouse effect and ancient loss of whatever near-surface water might have been there when it formed.

Mars, with its white polar ice caps that expand and recede in concert with the 687-day cycle of Martian seasons, is an altogether different story. Astronomers have long speculated that the red planet might be a wet, living world. In the 1870s, during a particularly close orbital approach of Mars to Earth, Italian astronomer Giovanni Schiaparelli documented dark, linear features that he interpreted as natural, possibly water-bearing channels, or *canali* in Italian. When the English translation of his original descriptions erroneously called them canals, implying high-tech engineered structures, the idea of an intelligent, extinct Martian race took on a life of its own. Most notable of these Mars life enthusiasts was Harvard-trained astronomer Percival Lowell, who became obsessed by Schiaparelli's discoveries in the 1890s. He used his family's wealth to construct a private observatory in Flagstaff, Arizona, and there he devoted himself to the study of Mars. Employing a state-of-the-art twenty-four-inch telescope under the clear Arizona skies, he imagined that he could resolve a vast network of canals stretching from the presumably ice-covered poles to the desiccated Equator. In his immensely popular books, *Mars* (1895), *Mars and Its Canals* (1905), and *Mars as the Abode of Life* (1908), Lowell describes the last desperate technological masterpiece of a vanished, water-starved race.

Lowell's colorful imaginings fueled a wave of science fiction stories and novels (including H. G. Wells's 1898 classic, *The War of the Worlds*) but did little to convince the scientific community that Mars is wet, much less living. In spite of more than a century of subsequent studies employing larger and larger telescopes, supplemented by a flurry of sophisticated Mars flyby missions (commencing with Mariner 4 in 1965), orbiters (Mariner 9 was the first in 1971), and landers (starting with Viking in 1976), definitive evidence for Martian water repositories and their significant extent proved elusive. The presence of water ice in the northern polar regions was finally, unambiguously documented by the Viking missions with spectral measurements in the late 1970s, but only since 2000, with an arsenal of instruments on the latest generation of satellites, coupled with scraping tools on the Phoenix lander and the Spirit and Opportunity rovers, has the true vast extent of water and the nature of its repositories on Mars been confirmed.

Today most of Mars's water occurs as subsurface permafrost and perhaps as groundwater in deeper warm zones—potentially huge repositories that remain hidden from the dry outermost layer. Hints of the extent of this subsurface water were provided in 2002 by the Mars Odyssey spacecraft, which carried a sophisticated neutron spectrometer. When cosmic rays bombard Mars's surface, they can dislodge neutrons from hydrogen-rich (that is, water-bearing) deposits. The spectrometer was designed to detect such neutrons as they spray out from a wide swath of the Martian surface, from the Equator to high latitudes. However, these intriguing results raised as many questions as they answered, for the exact form of water—liquid, ice, or mineral bound—could not be determined.

In 2007 NASA's Mars Reconnaissance Orbiter, using ground-penetrating radar, provided a much higher-resolution picture of this buried water. These pioneering measurements detected glacier-size accumulations of ice in the mid-southern latitudes. More recently, the European Space Agency's Mars Express Orbiter employed a similar radar system to detect deep ice across a wider swath of the planet. Some areas near the south pole reveal ice-rich zones more than fifteen hundred feet deep. Indeed, Mars may hold a quantity of subsurface frozen water equivalent to a globe-spanning ocean hundreds of feet deep. So Earth's oceans may have once had a Martian cousin.

Water can also be revealed by the presence of distinctive rocks and minerals. NASA's Mars Express, Phoenix lander, and its plucky rovers, Spirit and Opportunity, found abundant complementary evidence in the form of minerals formed through water-rock interactions. Water-rich clay minerals turn out to be a common phenomenon in the near-surface environment of Mars, and they may represent much of the hydrogen-rich material that the neutron experiments observed several years earlier. Evaporite minerals characteristic of dried-up lakes or oceans are also common, as is opal—a poorly crystallized variety of quartz that typically forms as hot water percolates through sediments.

As planetary scientists examine the red planet with new eyes, they also see more and more evidence that water once flowed freely across the sculpted Martian surface. High-resolution photos reveal ancient river valleys and gullies strewn with boulders, teardrop-shaped islands, slumps, and braided stream channels. These landforms cut through layers of sediment that appear to have been laid down in shallow lakes or seas. Indeed, the beachlike terraces that girdle the northern hemisphere of Mars imply that northern-latitude oceans may once have covered more than a third of the Martian surface. If so, then cooler Mars may have been a blue, life-giving planet many millions of years before Earth.

And then there's the Moon—a key to understanding the history of water on its larger companion, Earth. The Moon is bone-dry by conventional wisdom (actually drier than bone, which retains a significant water component even when baked in the desert sun). Multiple lines of evidence point to this aridity: Earth-based telescopes reveal no characteristic infrared absorption; Moon rocks from all six Apollo landing sites held no detectable traces of water (at least by 1970 analytical standards); and the finding of unrusted iron metal after four billion years on the lunar surface would seem to preclude even a trace of corrosive water.

It's a funny thing about conventional wisdom, though. Eventually someone will challenge what everyone else knows to be true, and once in a while something really interesting will be found. In 1994 a single flyby of the Clementine mission produced radar measurements that were consistent with water ice, though many planetary scientists were unconvinced. Four years later the Lunar Prospector employed neutron spectroscopy to detect a significant concentration of hydrogen atoms, and hence possibly water ice or water-containing minerals, near the poles. Still, many experts pointed to implanted hydrogen ions from the Sun's solar wind as a more likely source of the signal. Then in October 2009 NASA smashed the upper stage of an Atlas rocket into one of these craters (the Cabeus crater, near the southern lunar pole) and scrutinized the plume of impact debris for signs of H_2O. Sure enough, the flurry of dust incorporated a small but significant amount of the life-giving stuff—enough to renew interest in lunar water and its possible origins. Three back-to-back articles in *Science* that same October established that evidence for water on the Moon is now unambiguous.

Enter Erik Hauri and his colleagues at the Carnegie Institution. Using an ion microprobe—a highly sensitive instrument that hadn't been available to the first generation of scientists who studied the Apollo samples—Hauri's team has revisited the colorful glass beads of the sort I studied in my first geology job, picking out Moon specks, back in 1976. Other scientists had examined the glass beads for signs of water decades earlier, but their detection capacities were no match for the ion microprobe's ability to resolve measurements at the scale of a millionth of an inch. Hauri and his coworkers polished a variety of glass beads so that their round cross sections were revealed in the ion probe. The beads' outer rims proved to be very dry, with only a few parts per million water, but the cores of the largest beads have as much as a hundred parts per million. Over billions of years, most of the glass beads' original water has evaporated to space, more from the outsides than from the cores. However, based on the significant amount of remaining water deep inside the beads, Hauri and his colleagues calculate that the original water content of the Moon's magma may have been as high as 750 parts per million—a lot of water, comparable to many volcanic rocks on Earth, and more than enough to drive surface volcanism that would have dispersed magma in explosive eruptions billions of years ago.

If that much water powered volcanoes in the Moon's past, then a great deal of water must still be locked somewhere in the Moon's frozen interior. And since the Moon formed primarily by Theia's wholesale excavation of Earth's primordial mantle, our planet's deep interior likely holds prodigious amounts of unseen water as well.

The Visible Water Cycle

However much water we end up finding on Mars or the Moon (and it now appears there's a lot), Earth remains our Solar System's one and only water world. The story of Earth's water—how much there is, what form it assumes, where it resides, and how it moves—is more than a little complicated. Until as recently as the 1990s, the oceans were thought to be by far the largest water repository, storing about 96 percent of Earth's accessible inventory. Ice caps and glaciers, with about 3 percent today (and perhaps no more than 5 or 6 percent even at ice age peaks of glacial advance), come in at a distant second. Groundwater (all the near-subsurface H_2O, both in well-defined aquifers and as more widely dispersed stores) accounts for 1 percent, while all the lakes, rivers, streams, ponds, and the atmosphere combined represent no more than about a hundredth of a percent of Earth's near-surface water supply.

All this water is in constant motion, shifting from one repository to another on a scale of days to millions of years. The dynamic, life-sustaining water cycle represents the most obvious source of change on our perpetually changing planet. Imagine the possible perambulations of a single water molecule—a molecule formed from an oxygen atom and two hydrogen atoms that have existed for many billions of years. Begin with our molecule in the mighty Pacific Ocean, where most of Earth's near-surface water molecules spend most of their time. A great ocean river of cold water, the California Current, sweeps that molecule from near Alaska southward along the California coast to Baja and the Equator. As the surrounding water warms and rises, the molecule reaches near the ocean surface and begins an epic clockwise journey around the North Pacific—first the North Equatorial Current that flows west to curve past Japan, then the North Pacific Current heading east to North America. As our molecule once again moves close to California, it happens to rise to the sunlit ocean surface and evaporates into the atmosphere, where clouds are beginning to form.

Prevailing winds sweep the thickening mass of rain clouds eastward, across the desert Southwest, into the elevated terrain of the Rocky Mountains. As the clouds rise into higher, cooler elevations, the rain begins to fall. Our molecule eventually descends to Earth as part of a raindrop; it follows a sinuous path from rivulet, to creek, to stream, to a swollen river that overflows its bank. To this point the water molecule's movements have been swift—a year or two to cycle the entire Pacific Ocean, a day or two to join the clouds and fall as rain, a week or so to flow across the hilly terrain. But as it soaks deeper into the ground and merges with a vast, hidden aquifer, the molecule may spend many thousands of years creeping through the subsurface realm.

Here human actions alter nature's ancient rhythm, for water-hungry farmers pump out immense quantities of deep water to sustain agriculture in the semiarid Southwest. Aquifers, mined of their water at unsustainable rates, are thus drying up. Our molecule succumbs to this trend and finds itself back at the surface, splashing over a Texas cornfield, where it quickly evaporates back into the cloudless sky and continues its eastward journey.

This story cycles without end. Some molecules are temporarily broken apart into hydronium and hydroxyl ions, only to recombine into new water molecules with new atomic companions. Other molecules become frozen into thick Antarctic ice, where they will remain locked for millions of years. Still other water molecules undergo chemical reactions to become part of clay minerals in the soil.

Life has become an integral part of the water cycle as well. Plants take up water molecules and carbon dioxide and combine them, in the Sun-driven process of photosynthesis, to manufacture roots, stems, leaves, and fruit. And when those nutrient-rich plant tissues are eaten by animals and broken down by the metabolic miracle of respiration, the waste products, exhaled with every breath we take, are reassembled molecules of carbon dioxide and water.

The Deep Water Cycle

In the mid-1980s Earth scientists began to think in earnest about water at a global scale, for the near-surface water cycle can't be the whole story. Because we know that magmas originating tens or hundreds of miles down hold enough water to cause explosive volcanism, we can assume that silicate minerals crystallized deep inside our planet must somehow trap H_2O. There must be a deep, hidden part of the water cycle that could tell us much about how and when Earth became the ocean-bathed planet it is today.

The experimental approach to deep water has focused on the possibility that the most common of minerals—olivine, pyroxene, garnet, and their denser deep-Earth variants—may be able to incorporate a small amount of water at mantle conditions. The study of water in "nominally anhydrous" minerals, which became a major focus of high-pressure mineralogy in the 1990s, yielded astonishing results. It turns out that at high pressure and temperature, it's relatively easy for some minerals to incorporate lots of hydrogen atoms, which are the mineralogical equivalent of water (because hydrogen atoms combine with oxygen in these minerals). Minerals that are invariably dry in the cooler, low-pressure environments of the shallow crust—where explosive volcanism releases any water—can become rather wet in the deep mantle.

In principle, the experimental strategy is pretty straightforward. Take a sample of olivine or pyroxene, add water, heat while squeezing, and see where the water goes. In practice, it's not so easy. In order to reproduce Earth's deeper mantle conditions, the sample must be pressurized to hundreds of thousands of times the atmospheric pressure (equivalent to millions of pounds per square inch) and simultaneously heated to temperatures as high as 4,000 degrees Fahrenheit. To accomplish this daunting feat, scientists employ two complementary high-pressure approaches.

Some rely on massive, room-size metal presses that exert tons of pressure on a tiny sample—elaborate variations on Hat Yoder's pressure bomb from half a century ago. One oft-used experimental assembly involves four nested stages like Russian dolls: each stage surrounds and hugs the next, focusing immense pressures onto a smaller and smaller volume. First, a giant pair of metal plates squeezes from above and below with awesome force that can reach thousands of tons. Those massive plates exert a viselike grip on the clever second stage, which consists of six curved, interlocking steel anvils—three above and three below—which in turn press uniformly from all sides against a third stage with a cube-shaped cluster of eight tungsten carbide anvils. The sample of powdered mineral plus water must be tightly encased inside the innermost fourth stage, often with a gold or platinum liner so that the reactants don't squirt out the sides. As if generating pressure isn't hard enough, the sample must also be baked with electrical heaters buried deep inside the sample holder, and the temperature has to be measured continuously with a delicate loop of special wire called a thermocouple.

Another popular experimental approach to simulating Earth's deep interior is the diamond anvil cell, which generates extreme pressures by squeezing two diamonds with flat tips together. First take two typical half-carat brilliant-cut gem diamonds, just like the stones in old wedding rings, and polish down the sharp points on the bottom to flat, circular surfaces a twentieth of an inch across—what will become the anvil faces. Then mount the diamonds in a precisely aligned metal vise, and between them place a thin piece of metal with a tiny hole punched out. Center the hole over the opposed diamond anvils, load it with water and mineral powder, and squeeze. A modest force on the diamonds creates a tremendous pressure because the anvils are so small and thus concentrate the force. Diamond anvil cells have sustained record high pressures equal to the three million atmospheres found in Earth's inner core. The beauty of the diamond anvil cell is that you can see your pressurized sample by looking through the transparent gems. A battery of analytical spectroscopic techniques can be brought to bear, and it's easy to heat the sample to mantle conditions with a high-powered laser, which can also shine through the transparent diamond anvils.

If everything goes well—if the desired pressures and temperatures are reached and maintained, if the thermocouple doesn't break, if the sample doesn't leak—then the tricky analytical tasks begin. Some water-bearing minerals like clays and micas are easy to recognize, but how do you measure a few parts per million of water in an otherwise dry sample? The ion probe is one option; its high sensitivity and spatial resolution led to Erik Hauri's discovery of trace amounts of water in lunar volcanic glass. Infrared spectroscopy, which can reveal characteristic bonds between oxygen and hydrogen, is another useful tool. Newly formed bonds between hydrogen and oxygen alter the way that infrared radiation interacts with a crystal—changes that can reveal water entering the mineral structure. Nevertheless, cautious colleagues (and wary rivals loath to be scooped) will always raise the possibility that experiments are flawed or analytical techniques too insensitive. A single fluid inclusion—a minute pocket of water too small to see with a microscope—can yield a false signal in such finicky measurements.

As in any new scientific endeavor, it took a while for these experiments to catch on, but the more scientists looked, the more minerals emerged as likely hosts for deep water. Olivine and pyroxene in the lower crust are fairly dry, holding no more than a hundredth of a percent water. But raise the pressure to mantle conditions of 100,000 atmospheres and the temperature to 2,000 degrees Fahrenheit, and olivine transforms to wadsleyite, which can incorporate a whopping 3 percent water. The corresponding Earth layer, the mantle's transition zone from about 255 to 410 miles deep, is one of the wettest places in the planet and may hold nine times all the water in the oceans. Minerals of the lower mantle are less waterlogged, but they make up for that in their huge volume—half of Earth's total—so the lower mantle is estimated to hold another sixteen times the water in Earth's oceans. Given the likelihood of other water-rich minerals and that Earth's iron core probably holds a lot of hydrogen as well, the deep interior may store more than eighty oceans' worth of water.

First Ocean

Conservative estimates place proto-Earth's original budget of volatiles at more than a hundred times modern levels. Indeed, one of the principal challenges in modeling the history of Earth's volatiles is figuring out how much was lost—and how it escaped.

Of some things we can be sure. From day one, volatiles were released prodigiously from the deep interior, as megavolcanoes pumped huge quantities of steam into a rapidly thickening atmosphere. In the first few million years of proto-Earth's existence, that first atmosphere may have been many times denser than that of the modern world. Water may have poured out onto the surface in liquid form, cooling the first rocks and forming wide, shallow seas within a few tens of millions of years.

And then the Big Thwack blasted it all away. Almost every molecule that had worked its way to the surface was lost to space, in what amounted to the pushing of a giant reset button. We have no reasonable estimates as to how much of Earth's store of nitrogen, water, and other volatiles was lost in that single event, but it was a lot. Dozens of smaller-scale impacts of hundred-mile-diameter rocks caused unimaginable disruptions for another five hundred million years, each one vaporizing a significant fraction of the oceans and further diminishing the volatile inventory.

Nonetheless, within a few million years following the Big Thwack, water vapor had again become a principal component of the primordial atmosphere, forming a global tempest of turbulent dark clouds, howling winds, shattering lightning, and unceasing torrential rain. The surface of the storm-lashed basalt crust cooled and hardened, as low-lying basins gradually filled, slowly forming the oceans. For a time the encroaching seas created a global sauna, as the thin veneer of surface water penetrated cracks and fissures, contacted the hot rocks below, and returned to the surface as gargantuan geysers of roaring steam and superheated water. Such intense water-rock interactions served to hasten crustal cooling, making way for deeper ponds, then lakes, then oceans.

The exact timing of the global ocean's formation is unknown, but tantalizing evidence has emerged in the form of Earth's oldest crystals. Some of the most ancient rocks on Earth are three-billion-year-old layers of sediments in the arid sheep-ranching terrain of Western Australia, known as the Jack Hills. The sand-size mineral and rock fragments that make up those sediments would have eroded from vanished rock formations that must have been much older. A tiny fraction of those sand grains, not more than one in a million, is made of the mineral zircon—zirconium silicate ($ZrSiO_4$), one of the toughest materials in nature.

Individual zircon grains, typically smaller than the period at the end of this sentence, first formed as a minor accessory mineral in igneous rocks. Imagine basalt solidifying out of a melt that holds only a trace of the element zirconium. Most of the chemical elements, whether rare or common, easily enter the crystal structures of pyroxene, olivine, and feldspar. But zirconium has no home in common minerals. Rather, it seeks its own kind and thus forms isolated, tiny zircon crystals.

Several factors act together to make these easily overlooked zircon crystals a unique source of insight about the earliest Earth. First, zircons can last almost forever (at least for all of Earth history). A single crystal of zircon can erode from one rock (perhaps the igneous host where it first crystallized), then become part of a sedimentary sandstone rock, and then erode again and again and again for billions of years. The same individual zircon grain can be recycled through a dozen different sedimentary rock formations.

Second, zircon crystals tell time because they readily incorporate the element uranium, which can make up 1 percent or more of their atoms. Radioactive uranium, with a half-life of about 4.5 billion years, is nature's ultimate stopwatch. Once a zircon crystal forms, its uranium atoms are locked in, and they begin to decay at a steady rate; half of them decay on average every 4.5 billion years, each one ultimately transforming into a stable atom of lead. The ratio of dwindling parental uranium atoms to growing daughter lead atoms provides an accurate estimate of the zircon crystal's age.

Finally, two of every three atoms in zircon are oxygen, which provides clues about the temperature of formation. Recall that one line of evidence regarding the Moon's formation was the distinctive ratio of oxygen's stable isotopes: Earth and the Moon have identical ratios of oxygen-16 to oxygen-18, which implies that they formed at a similar distance from the Sun. In a similar line of reasoning, the ratio of oxygen-16 to oxygen-18 in a zircon crystal points to the temperature at which it grew: heavier oxygen-18-enriched samples point to a cooler temperature of formation. For igneous rocks, this temperature can be a sensitive indicator of the water content of the magma in which zircon crystals grew, because water lowers the temperature at which crystals grow. What's more, isotope ratios from water near Earth's surface tend to be even more enriched in heavy oxygen, so zircon crystals with extremely high oxygen-18 contents have been interpreted as having interacted with surface water.

In this way, zircon crystals from Earth's earliest rocks can survive many cycles of erosion and deposition, while they preserve details of the age, temperature, and water content of their original environment. All that information gleaned from crystals barely large enough to see without a microscope!

The bottom line is that many individual zircon crystals from the Jack Hills of Australia are more than 4 billion years old, with one aged, ancient sand grain clocking in at a remarkable 4.4 billion years. That oldest zircon crystal—indeed, the oldest known surviving solid fragment of Earth—has a surprisingly heavy oxygen isotope composition. Some scientists conclude that 4.4 billion years ago, when Earth was only about 150 million years old, the surface was relatively cool and wet: hence there were oceans.

Other experts aren't so sure. They point out that zircon crystals can be incredibly complicated: that 4.4-billion-year-old grain, as well as virtually all its slightly younger companions from the Jack Hills, has an ancient crystal core. But detailed mapping of each individual crystal reveals concentric layers of younger zircon grown around old layers. It's not uncommon for a single grain to display a billion-year range of ages from core to rim, with correspondingly complex variations in oxygen isotopes. If the older core was altered during more recent pulses of crystal growth, then the true character of Earth's ancient surface might be obscured.

Whatever the eventual outcome of the zircon story, most experts agree that not much more than one hundred million years after the Big Thwack, Earth had become a brilliant blue water world with a mile-deep encircling ocean. From space it would have appeared as an ultramarine marble, swirled with white wisps of clouds to be sure, but predominantly a breathtaking blue. (The ocean's color arises from simple physics. The sunlight that bathes the surface comprises all the colors of the rainbow—reds, yellows, greens, and blues—but water absorbs the red end of the spectrum more easily, so our eyes perceive the predominance of the scattered blue wavelengths of light.)

And what of the land? Today the continents occupy almost a third of Earth's surface, but at the dawn time of our planet, during the hellish Hadean Eon, continents had not yet formed. The primordial blue ocean was broken only by isolated steaming volcanic islands that poked above the waves. Their cone-shaped contours and narrow, rubbly black beaches, randomly dotting the globe from the poles to the Equator, were the only features to break the watery monotony.

As we think back to Earth's earliest, globe-spanning ocean, we wonder what it was like. Was it hot? Probably at first, given the still-cooling global magma ocean beneath. Was it fresh or salty? Salt is perhaps the most distinctive property of modern ocean water, but it might seem reasonable to assume that Earth's first ocean started out fresh, with few dissolved chemicals, and only gradually achieved the saltiness we find today. On the contrary, recent evidence suggests that the hot early ocean quickly became far saltier than today. Common table salt, sodium chloride ($NaCl$), readily dissolves in hot water. Today about half of Earth's salt is tied up in landlocked salt domes and other evaporite deposits related to dried-up bodies of salt water. Most of this salt is sequestered in thick layers deep underground, but during

Earth's first half-billion years, there were no continents on which to harbor salt. Consequently, the salinity of the first ocean may have been as much as twice that of the modern world. Moreover, other elements dissolved in the warm ocean water—primarily iron, magnesium, and calcium that dominate basalt—would have been present in higher concentrations as well.

Scientists also wonder if the Hadean ocean was acidic or basic. The most critical single factor controlling the ocean's pH and salinity is atmospheric carbon dioxide. By most accounts, the CO_2 content of the early atmosphere was thousands of times higher than today's value of a bit less than four hundred parts per million (though year by year we are rapidly approaching and soon will surpass that level). A lot more CO_2 in the Hadean air meant a lot more CO_2 in the water as well, and that must have had significant consequences for both pH and salinity. Carbon dioxide combines with rainwater to form carbonic acid, H_2CO_3. In the ocean, this carbonate partly dissociates to hydrogen ions, which form hydronium ions (the H_3O^+ of acids) plus bicarbonate (or HCO_3^-). That net addition of H^+ turns the oceans more acidic, perhaps to a pH as low as 5.5. Such acidic ocean conditions in turn probably accelerated the weathering of basalt and other rocks, adding even more solutes to an already salty ocean.

The Faint Sun Paradox

As if the detailed, sometimes conflicting stories of Earth's first ocean weren't controversial enough, there's one additional big wrinkle to contend with: according to increasingly sensitive astronomical observations and astrophysical calculations, stars like our Sun undergo a slow, inexorable brightening over their lifetimes. By these estimates, the youthful Sun of 4.4 billion years ago was 25 to 30 percent less bright than it is today. What's more, the Sun would have remained uncomfortably faint for at least another 1.5 billion years. If today's Sun were suddenly to dim by that extreme amount, Earth would quickly enter a devastating icebox phase; the oceans would freeze solid from the poles to the Equator, and most life on Earth would die. Only the hardiest organisms, deep-subsurface microbial life and animals living in protected hydrothermal zones associated with volcanoes, could survive such a catastrophic climate change.

Given such a cooler early Sun, Earth must surely have quickly frozen over. And yet geological evidence for abundant surface water at least as far back as four billion years is unambiguous. Sediments from both shallow- and deep-water environments are common. Life began and thrived during that interval. How, then, could the early ocean have been liquid?

Certainly some of the heat deficiency caused by the much fainter Sun was compensated for by a much hotter Earth. Following the crusting over of the primordial magma ocean, there was still a lot of hot molten rock and volcanic activity to warm the surface. An ocean on such a planet would have been continuously heated from below, as the black crust slowly thickened and cooled.

The leading hypothesis to explain the faint Sun paradox points to an exaggerated greenhouse warming effect caused by extremely high atmospheric concentrations of carbon dioxide—perhaps more than ten times the pressure of our present atmosphere (the same high CO_2 concentrations that may have acidified the ocean and increased its salinity).

A second clever scenario posits that Earth, in its early black, then blue, phases, absorbed a far higher percentage of the Sun's energy than the surface does today. Today the oceans absorb more sunlight than the land—an effect possibly exaggerated long ago by the high concentrations of iron in the earliest oceans. That increased solar absorption was coupled with a probable dearth of light-scattering clouds; today, plant-generated particles and chemicals play a major role in nucleating clouds, but billions of years ago there were no plants to trigger cloud formation.

Yet another hypothesis places a large amount of the potent greenhouse gas methane in the early atmosphere. A curious consequence of a methane-rich atmosphere would have been chemical reactions high in the atmosphere, where ultraviolet radiation would have triggered the synthesis of a rich variety of organic molecules, including possible building blocks of life. Such organic molecules might have caused a thick, smoglike haze, transforming the blue Earth into a distinctly orange world, not unlike Saturn's big moon Titan. And so while we don't yet know the exact combination of factors, we have more than enough explanations as to how Earth kept well above the freezing point.

What we can say with confidence is that, once formed, this global ocean shaped the planet's outermost layers—in sculpting the land, in the evolution of the increasingly diverse mineral kingdom, and in the origin of the biosphere. Water still works its magic in every facet of our lives, as the concentrator of mineral wealth, as the principal agent for surface change, and as the medium for all life.

Chapter 5

Gray Earth

The First Granite Crust

Earth's Age: 200 to 500 million years

᠙ Earth today is a world of contrasts—one-third land, two-thirds water; seen from space, a mélange of blue, brown, green, and swirling white. Not so 4.4 billion years ago, when widely scattered, symmetrical volcanic cones of black basalt were the only meager parcels of dry land to poke above the blue monotony of the shallow seas. All that was about to change with the invention of granite—the rugged foundation stone of the continents.

Earth's story is a saga of differentiation—of the separation and concentration of elements into new rocks and minerals, into continents and seas, and ultimately into life. Time and time again this theme has played out. The inner rocky planets—Mercury, Venus, Earth, and Mars—formed when intense pulses of solar wind separated hydrogen and helium from the heavier big six elements, sweeping the lighter gaseous elements outward to the domain of the giant planets Jupiter, Saturn, Uranus, and Neptune. On Earth, dense molten iron settled to the center, as the metal core separated from the peridotite-rich mantle. Partial melting of peridotite produced basalt, a rock rich in silicon, calcium, and aluminum, which separated from peridotite to form Earth's first thin, black crust. As basalt erupted explosively onto the surface, water and other volatiles separated from the basaltic magma to form the first oceans and atmosphere. Each heat-driven step separated and concentrated elements; each step led to an increasingly layered, differentiated planet.

The rise of continents was yet another important step in the differentiation of Earth. As the outer crustal layers of basalt cooled and hardened, they formed a lidlike, heat-trapping cover to the partially molten mantle beneath. Basalt, reheated from below, began to melt at relatively low temperatures, especially in the presence of water—as chilly as 1,200 degrees Fahrenheit. As the temperature increased, so did the percent of basalt melting—first 5 percent, then 10 percent, eventually up to 25 percent melt. In an echo of peridotite melting, the resulting magma was sharply different in composition from its host basaltic rock. Most notably, this new melt was much richer in silicon, with a significantly enhanced component of sodium and potassium as well. Water, too, concentrated in this hot fluid, as did dozens of rare trace elements—beryllium, lithium, uranium, zircon, tantalum, and many more. This new silicon-rich magma was much less dense than its parent basalt, so it inevitably pushed its way toward the surface, forming the first granite.

Most granites host a simple mineralogy of four different species. Clear, colorless crystals of quartz—pure silicon oxide—abound in granite; their tough grains would erode to produce Earth's first white sandy beaches. Two kinds of feldspar, one rich in potassium and the other in sodium, gave Earth's earliest granites their monotonous grayish-white color. And sprinkled in every granite is a fourth, darker, iron-bearing mineral—sometimes blocky pyroxene, sometimes sheetlike mica, sometimes elongate amphibole. The next time you see a polished granite countertop or bathroom fixture, take a look for this simple suite of four minerals.

The presence of rarer elements often leads to scattered tinier grains of additional minerals, such as zircon, for example, which concentrates zirconium. Recall from the last chapter that minute, gemmy red zircon crystals extracted from the remote Jack Hills of Australia provide hints of an early ocean 4.4 billion years ago. Those same crystals, which appear to have formed under relatively cool, wet conditions, may also point to the beginnings of granite formation at that early stage. Not only do the Jack Hills zircons bear the distinctive heavy oxygen isotope signature of a cool and wet origin, but a few 4-billion-year-old crystals also hold inclusions of quartz—a mineral rarely produced before the advent of granite. Some experts suggest that these old, cool, quartz-bearing zircon crystals are the last surviving remnants of the earliest granite crust.

With the origin of granite, we see for the first time a significant divergence of Earth's mineral evolution from that of some of its planetary neighbors. Granite formation requires abundant basalt near the planetary surface as well as intense internal heat to remelt it. The smaller planets Mars and Mercury, as well as Earth's Moon, are girdled by the necessary basaltic veneer, but they are too small to make much granite. They lack the necessary internal heat. Small volumes of granite were undoubtedly generated on these worlds, but nothing like the deeply rooted granite continents of Earth.

Buoyancy

Earth's primordial crust of black basalt, softened by heat from below and with a uniform density about three times that of water, never could support much topography. A few volcanic edifices may have soared a mile or two above the mean, enough for scattered black islands to rise above the sea, but there were no great mountain ranges, nor any deep ocean basins before the rise of continents. Granite, with a significantly lower average density (about 2.7 times that of water), changed that dynamic. Granite inevitably floats on basalt and perido-tite; it piles up in great mounds, rising miles above the surface like an iceberg floating on water.

Ice, which is about 10 percent less dense than water, provides a familiar analogue. Because of this density difference, about 10 percent of an iceberg's volume sticks out of the water. A jagged two-hundred-foot-tall iceberg will typically expose thirty or more feet above the surface; hence "the tip of the iceberg." By the same token, granite is 10 percent less dense than the basalt on which it floats. As the Earth's partially melted basalt crust generated layer upon layer of granite, iceberglike protrusions began to form. A mile-thick granite body might have produced a small mound that projected almost a thousand feet above the average level of basaltic crust. But over time, accumulating masses of granite crust reached thicknesses of many miles; accordingly, deeply rooted continental landmasses rose higher and higher above the oceans, with some mountain ranges soaring miles above the water's surface. Today's Rocky Mountain chain in the American West, with granite roots as much as forty miles deep, boasts numerous peaks over fourteen thousand feet. This grand spine of the North American continent stands tall as testimony to the buoyancy of granite.

In 1970, when I took my first geology course at MIT, the power of buoyancy in driving geological change was still textbook orthodoxy. (We used the 1965 edition of British geologist Arthur Holmes's richly illustrated classic, *Principles of Physical Geology*.) "Isostasy," it was called. The driving force of "vertical tectonics" was "isostatic readjust-ment." A neat woodcut, virtually unchanged from nineteenth-century geology texts, showed a line of rectangular wooden blocks of different heights floating in water. Taller blocks projected higher out of the water, just like a mountain. We learned how ocean basins had filled with thick layers of sediments, and how those sediments had melted to form more granite bodies. We learned how mountains subsequently rose from those buoyant granitic cores. It all made perfect sense at the time, and it's still the leading hypothesis for how Earth's earliest crust formed more than four billion years ago.

Early in Earth's history, perhaps even within the first two hundred million years, modest landmasses of buoyant gray granite must have begun to form above hot spots, as deep accumulations of basalt were partially melted. In those early times vertical tectonics and isostasy must have prevailed, just as Arthur Holmes taught. Those first isolated continental bits of granite were utterly barren and windswept and battered by intense waves. Eroded rock fragments of quartz slowly accumulated to form meager sandy beaches, while feldspars weathered to thin layers of clay-rich soils. The first granite islands were isolated, modest in size, and low in profile; they gave no hint to the scale of the continents to come.

Impact Redux?

So how did early Earth shift from a volcano-dotted basaltic world to a planet with wide, gray granitic continents? How did the first few lonely granite islands expand to the hemisphere-spanning landmasses we see today? Earth scientists have never been shy in devising hypotheses. One of the more intriguing ideas posits a continent-forming sequence triggered by that familiar agent of chance: the careening asteroid.

For a billion years after Theia's obliteration and the Moon's formation, huge impacts occurred from time to time. This is indisputable. Experts estimate that dozens of large asteroids up to a hundred miles across—wandering remnants of the original planet-forming era— must have collided with Earth during its formative aeons. Imagine a scenario four billion years ago, as a hot plume of softened rock ascends beneath a young ocean crust. Many dozens, if not hundreds, of such plumes must have risen from Earth's deep interior to transfer inner heat by the efficient process of convection. Great volcanoes spewing basaltic lavas have erupted above each plume, even as remelting of basalt crust generated a granitic component that thickened the land.

Then comes catastrophe: a thirty-mile-diameter asteroid slams into the volcanic cluster, eradicating every trace of land within a three-hundred-mile radius. The impact produces a giant, bowl-shaped magma lake while showering the adjacent surface with blobs of sticky molten lava and shattered rock. This cosmic insult blocks the mantle plume, which has to find a new path to the surface.

According to this clever scenario, the postimpact plume changes path to poke up underneath a minicontinent, with basaltic roots and a growing granite veneer. Once it is positioned below such a thick, heat-trapping lid of basalt, the new heat source produces new pulses of granite in abundance, thus expanding and thickening the land.

This untestable story is probably part of Earth's earliest continent-forming narrative. A billion years of vertical tectonics enhanced by asteroid collisions would have generated an ever-increasing inventory of volcanic ocean islands with intermixed basalt and granite cores. Gradually land arose from the sea. By four billion years ago, large islands, distributed at random across the globe, may have occupied a modest percentage of Earth's surface.

But then along came plate tectonics, and Earth's near-surface evolution shifted into high gear.

Drifting Continents

The discovery of plate tectonics as Earth's dominant geological process is itself a story spanning most of modern science. Though anticipated by at least four centuries of observations, the idea that entire continents could somehow migrate across Earth's surface was dim and heretical at first, coming into sharp focus and widespread acceptance only in a furious international rush of discoveries in the 1960s. But once the flood of evidence began to pile up, the Earth sciences experienced one of the most rapid paradigm shifts in the history of science. Indeed, within five years of my undergraduate tenure at MIT, by the mid-1970s, every geology textbook had to be completely rewritten as the old orthodoxy of vertical tectonics was all but expunged.

In retrospect, some of the evidence against vertical tectonics should have been obvious. As tall as the Rocky Mountains stand today, they are dwarfed by the seven-mile height of Mount Everest and the mighty Himalayan range. Likewise, in contrast to the two-mile average depth of the oceans, Earth's deepest ocean trench, located off the Mariana Islands in the South Pacific, plunges to an astonishing seven miles. Such topographic extremes could not possibly be sustained in an isostatic world. Vertical tectonics couldn't be the whole story.

Subtle hints of lateral tectonics—the role of sideways motions in Earth's geological evolution—came with the first accurate maps of the New World's coastline. By the early 1600s, the striking conformity between the eastern coastline of the Americas and the western shores of Europe and Africa was plain to see. The same sinuous shape, the same embayments and bumps, the rounded contours of extreme southwestern Africa and the matching suggestive eastward curl of South America's tip—all pointed to some ancient jigsaw-puzzle-like fit.

Several bizarre hypotheses attempted to explain the tantalizing transatlantic continental match. Astronomer William Henry Pickering of Harvard University, who supported George Darwin's theory of fission origin of the Moon (as a molten blob hurled into space from a rapidly spinning Earth), posited that simultaneously as the Moon was ripped from the Pacific Ocean, on the opposite side of Earth the Atlantic Ocean opened wide. Others saw the hand of God in the great Atlantic S. Perhaps the Atlantic coastlines were the shores of

Noah's mighty flood, which had been unleashed a few thousand years ago to create the great ocean and "divide the lands."

Systematic geological surveys might have helped resolve the question, but four hundred years ago geology had not even been named, much less pursued in any systematic way. Mining and agriculture, the driving economic forces behind the earliest geological surveys of the late eighteenth century, were strictly state and national affairs. Little effort was made to match up geological formations across political boundaries; nor were the riches of one principality seen as linked in any coherent way to those of any other. Gold was quite literally where you found it. In such a nationalistic mapping milieu, matching up geological features across the great expanse of the Atlantic Ocean was hardly a priority.

The first detailed transatlantic geological comparisons were undertaken by an unlikely scholar, meteorologist Alfred Wegener, who spent much of his career in the Arctic. (He died at age fifty during a heroic winter rescue mission on the frigid Greenland ice sheet.) Though his professional life was devoted primarily to studying the origins of weather, his most memorable and lasting work related to what he called "continental drift," an early and much disparaged contribution to lateral tectonics. The inspiration for this odd geological digression came during World War I, when he served as a reserve lieutenant in the German army. Shot through the neck during the Belgian campaign, Wegener was relieved of frontline duty and permitted to devote his convalescence to study.

Wegener, like many of his predecessors, was struck by the apparent fit of continents across the Atlantic Ocean, though many scientists had dismissed the match as coincidence. Wegener cast a wider field of view and realized that similar fits could be seen in the varied coastlines of East Africa, Antarctica, India, and Australia. Indeed, all of Earth's continents could be elegantly clustered together to make one supercontinent, which he dubbed Pangaea (from the Greek for "all lands"). Wegener and a handful of like-minded supporters also cited evidence from recently published geological surveys of coastal regions of Europe, Africa, and the Americas—treatises that revealed tantalizing correlations across the wide expanse of the Atlantic. Great mining districts, such as the extensive gold and diamond reserves of Brazil and South Africa, appear as a single large deposit when the continents are juxtaposed. Similarly, rock layers bearing the distinctive fossil fern *Glossopteris* and the extinct reptile *Mesosaurus* line up almost exactly. Such detailed geological and paleontological correlations could not be simple coincidence, he argued.

Wegener's continental drift hypothesis first appeared in print in 1915. Three subsequent German editions, each more detailed than the last, as well as a 1924 English translation entitled *The Origins of the Continents and Oceans* and many other editions, followed. New data poured in to support the idea that continents had once been joined together. In 1917 a committee of paleontologists cataloged more than a dozen instances of distinctive fossil-bearing strata matching up across the oceans—data they interpreted as requiring some sort of ancient land bridges. South African geologist James Du Toit, who was especially enamored of Wegener's ideas, obtained a grant from the Carnegie Institution to visit eastern South America. He recorded more examples of transoceanic matchups: striking instances of identical minerals, rocks, and fossils.

Yet despite the accumulating data for continental alignment, the Earth science community was unmoved. Lacking a plausible mechanism for continental-scale wanderings, many geologists were openly contemptuous of Wegener's conjectures. They were bolstered in these criticisms by Newton's laws of motion, which mandate that big continents can't wander across the globe without a correspondingly epic force to move them. Until a force of global scale could be invoked, continental drift would be viewed as little more than a crackpot idea by a geological amateur. Cambridge physicist Harold Jeffreys summed up the British viewpoint in 1923: "The physical causes that Wegener offers are ridiculously inadequate." Geologists in America were equally unconvinced. Rollin T. Chamberlin of the University of Chicago's geology department blasted continental drift at a 1926 symposium: "Wegener's hypothesis in general is of the foot-loose type, in that it takes considerable liberty with our globe, and is less bound by restrictions or tied down by ugly facts than most of its rival theories. . . . If we are to believe Wegener's hypothesis we must forget everything that has been learned in the last 70 years and start all over."

Even so, a few Earth scientists were sufficiently intrigued by the findings of Wegener and his supporters to devise novel mechanisms for continental shifts. One school of thought posited that Earth is shrinking, perhaps by cooling or by collapse of gas-filled voids in the deep interior, and thus portions of the surface must gradually fall inward like a broken archway. In this untenable model, the continents once boasted a continuous expanse of land from the western coasts of the Americas all the way to the eastern coasts of Africa and Asia. Today's Atlantic Ocean was viewed as a giant archway of land that has collapsed into the mantle. Basic Euclidian geometry foiled this shrinking Earth model: a simple archway can collapse, but transfer that idea onto a sphere and there's no way a continental volume covering the area of the Atlantic Ocean could collapse into anything.

Another group proposed the antithetical view that Earth has been expanding, inflating like a balloon over geological time. Once upon a time, there was only continental crust, which has cracked and split apart as the planet inflated (by some accounts from the generation of deep, hot expanding gases). Indeed, if you play an imaginary videotape of a supposedly expanding Earth backward, you can arrive at a state where all the continents slide neatly together to cover a sphere that is about three-fifths the diameter of the modern Earth. Lacking any other widely accepted formation mechanism for the Atlantic, this hypothesis persisted in some geological circles from the 1920s to as late as the 1960s, when a compelling new idea took its place.

The Hidden Mountains

Fast-forward to the post–World War II years, a time of tremendous technological innovation and optimism in science. Two developments in antisubmarine warfare, both declassified and adopted by oceanographers in the 1950s, led to transformative discoveries about the dynamic Earth.

Sonar, which uses sound waves to measure distance and direction, is a century-old technology familiar to anyone who has watched Hollywood submarine movies. You hear a PING, which is answered a short while later by a softer, echoing *ping*. A sound wave has bounced off the solid hull of a submarine. (The effect on the viewer depends on whether the movie's point of view is the hunter or the hunted.) "PING . . . . . . *ping*," "PING . . . . *ping*," "PING . . *ping*": the echoes come faster as the submarine's location is pinpointed. The tense music builds; depth charges are released.

The exact same technology can be put to scientific use to study ocean depth and thus ocean-floor topography. Even the deepest ocean valleys and trenches can be plumbed with sound waves. As early as the 1870s, British scientists employed crude deep-water soundings aboard HMS *Challenger* and reported hints of great mountains on the floor of the mid-Atlantic—a tantalizing result that some contemporary romantics associated with the lost continent of Atlantis. Primitive echo sounding technology, first developed for iceberg detection after the 1912 *Titanic* disaster, enjoyed rapid improvement during World War I, as German submarines began to prowl. The 1920s saw the first systematic application of sonar to mapping the ocean floor and the rapid realization that great mountain ranges lay hidden beneath all the Earth's oceans. However, the geological implications of these pioneering ocean surveys were little noted, and oceanographic efforts were largely curtailed by the Great Depression and the looming Second World War.

Following the war, oceanographers were armed with a new generation of high-sensitivity sonar detectors that could not only map the topography of the entire ocean floor but also detect reflected sound waves from deeper rock layers. General features of the Atlantic Ocean floor were easily confirmed. For example, continental shelves deepen gradually as you move away from most Atlantic coastlines, for distances up to hundreds of miles. The edges of these continental shelves are marked by a sudden drop-off to an abyssal plain two miles deep and a thousand miles across—a feature much wider and flatter than any on dry land. And the ocean is bisected by an extensive mountain range, the Mid-Atlantic Ridge.

All that was in accord with earlier discoveries, but the thickness of the ocean crust came as a huge surprise. Geologists had predicted that the oceans would be less deeply rooted than land, with ocean crust gradually thinning away from shore. What they found, instead of this gradual transition, was a remarkably abrupt contrast from thick to thin. Unlike the tens of miles of crustal rocks beneath the continents, ocean crust was only about five or six miles thick: the sharp transition occurred right at the drop-off at the edge of the continental shelf. Such a narrow demarcation between continents and oceans was at odds with isostatic models.

Year after year, back and forth across the wide ocean the scientists sailed, hundreds of times. Each and every crossing yielded the same result. A vast mountain chain more than twenty thousand miles long, the largest on Earth, lay beneath the waves, precisely bisecting the Atlantic. The same sweeping curves of the continental shorelines were mimicked in the hidden Mid-Atlantic Ridge crest. What's more, if the edge of the continents was taken as the sharp underwater drop-off to the abyssal plain (as opposed to the shifting sandy shoreline), then the match between continents was uncanny, as if a broken china plate were fit neatly back together. No longer could science dismiss the conformity of coastlines as mere coincidence.

As scientists completed more traverses of the Atlantic and compared more details, new patterns emerged. The Mid-Atlantic Ridge was no ordinary mountain range. On land, most mountain chains have a line of the highest peaks down their axis, but at the very centerline of the Mid-Atlantic Ridge was a wide trough about twenty miles wide and more than a mile deeper than the adjacent peaks to the east or west—a feature we now call a rift valley. What's more, the ridge and its rift valley didn't follow a smooth and continuous curve from north to south. Rather, the rift valley was repeatedly offset, a hundred miles or more to the east or to the west, by sharply defined transform faults, places where the crust was broken and displaced, giving the entire ridge a jagged and broken appearance. What was going on?

Such suggestive findings could have easily been buried in the avalanche of brilliant postwar scientific discoveries. In one sense, they were just more data. But the lead investigators on the ocean-floor project were no ordinary publicists. Bruce Heezen and Marie Tharp, marine geophysicists at Columbia University's Lamont Geological Observatory, devised a new, dramatic topographic map of Earth's surface. As on other topographic maps, they represented continental elevations with colors—higher elevations grading from greens and yellows to browns and ultimately white at the highest, snowcapped-mountain elevations. Great mountain ranges—the Himalayas, the Andes, the Alps—stood out clearly. Heezen and Tharp's artistic innovation was to highlight the immense subsurface mountain ranges in exactly the same way, albeit in varying shades and hues of blue—a technique that made the Mid-Atlantic Ridge and other deep-sea features stand out as monumental on a global scale. And by centering their exquisite map on the Atlantic, they highlighted the identical shapes of the coastlines and the ridge in an unmistakable way. By the 1960s, Heezen and Tharp's map had achieved iconic status. Whatever the cause of this parallelism, the fact that there was some genetic link was obvious to all.

(This story of Bruce Heezen—pronounced "HAY-zen"—and his widely lauded contribution has special meaning to me and my career, for when I arrived at MIT in the fall of 1966, I was surprised to find that even the most senior geology faculty members were quite respectful and eager to shake my hand. Distinguished pedigrees—even of erroneous homonymous variety—are not without some advantage in science.)

The Expanding Sea

With the revelation of the Mid-Atlantic Ridge, and the discovery of similar subsurface volcanic ridges in the East Pacific and Indian oceans, scientists tackled the possibility of lateral continental motions with renewed intensity. To be sure, continents aren't drifting aimlessly, as Wegener's coinage might suggest, so geologists looked for some hidden force that could dramatically rearrange Earth's surface.

Discovery followed discovery, as new data kept arriving to confound the experts. In 1956 Heezen and his Lamont boss, seismologist Maurice Ewing, documented a remarkable association between the position of the central rift valley of the Mid-Atlantic Ridge and a 34,000-mile-long pattern of moderate ocean-floor earthquakes extending around the globe. Somehow the rift valleys and earthquakes are related, so ridges must be dynamic, changeable features.

Ocean floor rocks also surprised many geologists, who had predicted that the Mid-Atlantic Ridge was a typical mountain range capped by resistant marine limestone, just like the Canadian Rockies. But extensive dredging along the ridge, coupled with observations of the Atlantic's many islands, produced nothing but basalt, and relatively young basalt at that. It turns out that, other than a veneer of soft sediments, the ocean's crust is made almost entirely of volcanic basalt. From east to west, spanning more than twenty-five hundred miles of ocean floor, basalt forms the pavement.

What's more, careful dating based on the steady decay rates of radioactive elements reveals a simple pattern in the ages of these rocks. Basalt collected from the rift valley right at the center of the Mid-Atlantic Ridge is newly minted, less than a million years old. The farther you get from the rift valley, east or west, the older the basalt is, until the rocks near the continental margins are more than one hundred million years old. Why should rocks at the center of the ocean be young, while those on the outskirts are so much older? One logical conclusion is that the Mid-Atlantic Ridge is a line of volcanoes that are spewing out new basaltic crust. But where did the much older rocks at the edges of the ocean basin come from?

Key data, the smoking gun of plate tectonics, came from a second submarine-hunting technology called the magnetometer. World War II submarines are big hunks of iron-rich alloys, so they are magnetic. Thanks to the development of magnetometers, submarine-hunting airplanes could fly over the ocean surface and pick up the magnetic anomaly of a nearby enemy submarine. Following the great conflict, geophysicists invented new types of magnetometers with greatly increased sensitivity to small changes in the magnetic field. They adapted these instruments to be towed behind research ships, just above the sea bottom.

Their target was ocean-floor basalt, which carries a weak magnetic signal in the form of minute crystals of the iron mineral magnetite. Earth's magnetic field is known to vary slightly from year to year, in what is called secular variation. As basalt magma cools, these crystals freeze in the direction of Earth's magnetic field like tiny compass needles. Ocean floor basalt thus preserves the orientation of Earth's magnetic field on the exact date when the rock hardened. The thriving field of paleomagnetism studies these invisible magnetic force fields that are locked into basalt and other rocks. (On dry land, a hodgepodge of magnetic signals, the consequence of folding, faulting, and other geological contortions of continental crust over time, confuses such patterns.)

Starting in the early 1950s, oceanographers deployed magnetometers close to the seafloor, sweeping them on long transects across ocean ridges. They hoped that their paleomagnetic measurements might yield a better picture of secular variation on the ocean floor. What they found instead was a bizarre magnetic pattern that was astonishingly regular and intricate. Close to the central rift valley in both the Atlantic and the Pacific, basalt displayed the normal magnetic orientation, faithfully pointing in the direction of the modern north magnetic pole. But several miles east or west of the rift valley, the magnetic signal flips a full 180 degrees: the north magnetic pole is almost exactly opposite its present position, where the south pole is supposed to be, and vice versa. Sail several more miles in either direction, and the magnetic field flips 180 degrees again to the correct orientation. Over and over, dozens of times, the magnetic field frozen in the rocks on any given transect is observed to flip.

Additional analysis revealed three key facts. First, the rocks with reversed magnetic field form long, narrow north-south trending bands that exactly parallel the ridges in both the Atlantic and the Pacific oceans. Where the central rift valley is broken, offset by transform faults, so too are the magnetic bands. Second, the pattern of these magnetic stripes is symmetrical about the ridge axes: sail east or west from the center, and the exact same sequence of normal and reverse stripes, some wider and some narrower, occurs. And third, radiometric dating of these basalts from ridge systems around the world confirms that each reversal occurred simultaneously over a narrow and precisely defined age. Magnetic reversals thus serve as a kind of ocean-floor time line.

Two logical, if mind-bending, conclusions followed. First, Earth's magnetic field is wildly variable: it flips 180 degrees on average every half-million years, and it has been doing so for at least the past 150 million years. The reasons for this fickle field are now understood in some detail. Our planet is a giant electromagnet; its magnetic field arises from swirling electrical currents in Earth's convecting fluid outer core. Heat drives this convection; dense hot liquid at the inner core boundary expands and rises, to be replaced by cooler, denser liquid that sinks from above. Geophysicists employ sophisticated computer models to show that Earth's rotation adds complicated, chaotic twists to the convection—motions that result in a flip in the magnetic field every half-million years or so. Earth's rotation also constrains the magnetic poles to spend most of their time aligned close to the stable rotation axis, but during periods of core instability, the magnetic field can wander widely and flip, perhaps in the span of a century or less.

The second conclusion is that midocean ridges produce new basaltic crust at the rate of an inch or more every year. Older basalt moves sideways, both eastward and westward, away from the ridge, as new lava takes its place. The ridge systems are thus grand two-directional conveyor belts, spewing out new ocean floor. Fresh basalt generated at the Mid-Atlantic Ridge is expanding the Atlantic, which grows as much as two inches wider every year—roughly thirty thousand years per mile of new ocean floor on average. Play the tape backward 150 million years, and the Atlantic didn't exist. Prior to that time, the Americas must have been joined to Europe and Africa, just as Alfred Wegener proposed.

One of the most influential presentations of this remarkable discovery appeared in 1961 in the *Geological Society of America Bulletin*. British geophysicist Ronald Mason and American electronics expert Arthur Raff of the Scripps Institution of Oceanography in California had collaborated for almost a decade, compiling exhaustive magnetic surveys of the ocean floor off the West Coast of North America. The centerpiece of their publication was a detailed magnetic map of the Juan de Fuca Ridge, a prominent ridge feature on the Pacific seafloor just a short day's sail from ports in Oregon, Washington State, and British Columbia.

The stark black and white map of Mason and Raff—white and black bands representing normal and reversed magnetic fields, respectively—displays dozens of north-south trending stripes. Uniformity prevails within large blocks of ocean crust, each hundreds of miles wide, each with a symmetrical pattern of stripes about a central rift valley. But between adjacent blocks the pattern is broken, offset by transform fault lines and skewed like a cubist painting. Analysis of offsets along one of these faults, the Mendocino Fracture Zone, reveals a remarkable lateral displacement of seven hundred miles. Epic internal processes must be at work to so disrupt Earth's crust.

With similar evidence piling up from ridge systems around the globe, geologists, geophysicists, and oceanographers began to talk to one another in a newly integrated effort. Correlations of ocean-floor topography, seismology, magnetism, and rock ages all pointed to the same conclusions. Ocean crust is being created around the world at ridge systems, which are dynamic zones of volcanic activity. The rate of seafloor spreading is recorded in the symmetrical patterns of magnetic stripes and the ages of basalts.

A flood of influential papers transformed the collective geological mind-set, and by the mid-1960s almost everyone was convinced of what had once been heresy: the continents are moving. The Atlantic Ocean has been growing wider every year for more than one hundred million years.

The Case of the Disappearing Crust

The early years of the plate tectonics revolution were a time of rapid discovery, changing paradigms, and equally puzzling new questions. One unanswered question stood out from the rest: How could a mile-wide swath of new basaltic crust be added every thirty thousand years along more than thirty thousand miles of midocean ridges in the Atlantic, Pacific, and Indian oceans? How did all that new crust fit? Unless Earth was growing larger—and for a brief interlude in the 1950s and early 1960s, a small but vocal group of geologists, including Bruce Heezen, did advocate the untenable expanding-Earth scenario—the old crust had to go *somewhere*.

Seismologists found the solution. In the cold war climate of the 1960s, nuclear weapons became the central focus (and primary source of funding) of seismology. Following the 1962 Cuban missile crisis, the United States and the Soviet Union agreed to a Limited Test Ban Treaty, which restricted nuclear weapons testing to underground detonations. Verification of the treaty relied on continuous seismic monitoring using an extensive (read: expensive) array of vibration-sensitive instruments deployed to the far reaches of the globe. The resulting World-Wide Standardized Seismograph Network (WWSSN) linked 120 stations to a central computer-processing center in Golden, Colorado, home of one branch of the United States Geological Survey. For the first time ever, it was possible to pinpoint the exact locations, depths, magnitudes, and motions of small earthquakes (and big explosions) anywhere on Earth.

The fringe benefits for Earth science were tremendous. Armed with their new tools, geophysicists could sense thousands of previously undetectable Earth movements and thus document previously unrecognized planetwide patterns of earthquakes. They found that almost all of Earth's sudden crustal motions occur along narrow lines of intense seismic activity—places like the midocean ridges. Many other quakes occur close to chains of volcanoes near the continental margins—for example, around the Pacific Ocean's notorious "ring of fire." These violence-prone regions of the Pacific rim, including the Philippines, Japan, Alaska, Chile, and other danger zones, formed a common pattern.

It had long been known that relatively shallow earthquakes (those from depths of a few miles or less) originate just offshore in the vicinity of deep trenches on the ocean floor, while deeper and deeper earthquakes, some originating more than a hundred miles down, occur farther and farther inland from the coast. The deepest known earthquakes commonly take place beneath chains of dangerous explosive volcanoes like Mount St. Helens and Mount Rainier in Washington State—volcanoes typically located at considerable distance inland.

By the late 1960s, new data from WWSSN clarified the details of the relationship among deep ocean trenches, earthquakes, and volcanoes. The distinctive pattern of quake depths increasing inland from the trenches painted a picture of huge slabs of ocean crust plunging down into the mantle, underneath the continents, along what were called subduction zones. Old basaltic crust, which is much colder and thus denser than the hot mantle, is literally swallowed up by Earth. As subducting basalt snags and buckles the adjacent crust downward, the deep ocean trenches form. For every square mile of new crust generated at ocean ridges, a square mile of old crust disappears at a subduction zone. The new exactly balances the old.

As if a veil had been lifted, the new science of plate tectonics came into sharp focus. Ocean ridges and subduction zones define the boundaries of about a dozen shifting plates, each of which is cold (compared with the deeper mantle), brittle (hence subject to earthquake fracturing), and only a few tens of miles thick but hundreds to thousands of miles wide. These rigid plates simply slide across hotter, softer mantle rocks. The Pacific Ocean's ring of fire defines one large plate, Antarctica and its surrounding seas another. The North and South American Plates extend westward from the Mid-Atlantic Ridge all the way to the Pacific coast of the Americas, while the Eurasian Plate extends eastward from the Mid-Atlantic Ridge to the Pacific coast of East Asia. The African Plate, stretching from the Mid-Atlantic Ridge on the west to the middle of the Indian Ocean on the east, displays an intriguing aspect of Earth's dynamic surface: the African continent is beginning to split apart as a new rift valley forms, marked by a string of lakes and active volcanoes, as well as high altitudes that regularly produce the fastest distance runners in the world. Someday Africa will become two plates, with a new expanding ocean in between.

As ocean ridges produce new plate material and subduction zones swallow up the old, the scenario is once again complicated by Euclid: Earth is a sphere. The geometry of plate growth and subduction on a sphere requires that some plates scrape against each other along jagged transform fault lines—hence the offset bands in Mason and Raff's famous magnetic map of the Juan de Fuca Ridge. The violent San Andreas Fault, which has triggered many memorable California earthquakes, is another such suture. Every day more stress builds along the fault, as the mighty North American Plate moves in a southeasterly direction relative to the mighty Pacific Plate. Every day these inexorable plate motions bring residents of Los Angeles and San Francisco closer to the next "big one."

So much for the simple geometry of plate tectonics. What of the epic forces that must power plate motions? What could cause entire continents to shift, scrape, and collide over hundreds of millions of years? The answer lies in Earth's inner heat. Earth is hot, while space is cold. The second law of thermodynamics, a sweeping core concept of the cosmos, states that heat always flows from hotter to cooler objects—heat must gradually disperse, must somehow find a way to even out.

Recall the three familiar mechanisms that facilitate the transfer of thermal energy. Every warm object transfers its heat to the surroundings in the form of infrared *radiation;* heat also moves, albeit much less efficiently, by direct contact or *conduction,* and by *convection,* when a fluid mass flows between hotter and cooler regions. Earth must obey the second law of thermodynamics. But how can heat move efficiently from searing core to cool crust? Rock and magma impede infrared radiation, while sluggish conduction is not much more efficient. So convection of softened, taffylike hot mantle rocks is the key.

Rocks at Earth's surface are hard, brittle materials, but deep inside the superheated pressure cooker that is the mantle, rocks soften like butter. Over millions of years, under the stresses of the deep interior, rocks deform, ooze, and flow. Hotter, more buoyant rocks gradually rise toward the surface, while cooler, denser rocks sink into the depths. Great convection cells, each thousands of miles across and hundreds of miles deep, overturn Earth's mantle in a majestic cycle hidden from view. The pace of these planetary shufflings is equally vast—it may take a hundred million years or more for a single revolution of the convection cell to complete itself.

At first, perhaps for more than a billion years, mantle convection beneath Earth's uniform basalt crust must have been a chaotic, swirling hodgepodge. Here and there, hotter melts of lower-density granite rose in disorganized pulses and plumes toward the surface, where they accumulated, disrupting the colder, denser basalt. Isolated dense chunks of that colder crust slowly sank into the interior, in a global-scale exchange of heat.

Over the next half-billion years, mantle churning became more organized. Dozens of smaller convection cells, each with rising plumes and sheets of magma and descending blocks of crust, consolidated into a handful of majestic cycles, each hundreds of miles deep and thousands of miles across. New, hot basaltic crust formed where these convection cells rose upward along growing seafloor ridges, while old, cold basaltic crust plunged into the mantle at a steep angle— subduction zones, in an Earth increasingly dominated by the new, transformative processes of plate tectonics. In cross section, Earth's turbulent outer layers might have looked like a collection of sideways whirlpools, each rotation lasting a hundred million years or more.

Then as now, Earth's evolving surface reflected the epic processes occurring far below. Great ridges of basaltic volcanoes grew above convection zones of rising magma. Gashlike trenches formed where old subducted crust plunged downward into the mantle, bending and buckling the adjacent ocean floor. Subduction also accelerated the all-important production of granite. As cold, wet, subducted basalt crust plunged deeper, swallowed back into Earth, it heated up and began to melt—not completely, but perhaps 20 or 30 percent. Those growing volumes of granitic magma rose to the surface, producing chains of gray volcanic islands hundreds of miles long. The stage was set to build the continents.

Revolution

Granite floats, basalt sinks: that's the key to the origins of continents. Magmas of granitic composition are much less dense than their parent basaltic rock, so these fresh melts slowly, inevitably rise to crystallize as near-surface rock masses or to erupt from volcanoes that spew layers of cinders and ash onto the surface. Over billions of years of Earth history, countless granite islands have formed by this continuous process.

Plate tectonics not only produced these granite-rooted island chains; it also assembled them into continents. The key lies in the simple fact that granite cannot subduct. The dense basalt on which it floats easily sinks into the mantle, but granite is like a buoyant cork. Once formed, it remains at the surface, conserved. As subduction produces more islands, the total area of granite irreversibly increases.

Imagine a subducting plate of ocean crust, dotted with unsinkable granite islands. The basalt subducts, but the islands don't. They must remain at the surface, to form a strip of land right above the subduction zone. As tens of millions of years pass, more and more granite islands pile up to form a wider and wider strip, even as new volumes of granite melt rise from the subducting slab to thicken and expand the growing continent. Islands accrete to form proto-continents, which accrete to form continents, just as our Solar System's chondrites once accreted to form planetesimals, planetesimals to form planets.

The epic cycle of plate tectonics transforms our world. Earth's thin, cold, brittle surface cracks and shifts like scum on a pot of boiling soup. New basaltic crust pours out of volcanic ridges that reveal where deep convection cells rise. Old crust is swallowed up at subduction zones that reveal where convection cells descend. Earth's most violent surface disruptions—the most intense earthquakes, the mightiest volcanoes—are but trivial, incidental blips compared with the vastly more energetic global-scale movements of the deep interior.

Plate tectonics also revolutionized the Earth sciences. In the prior dark ages of vertical tectonics, each geological discipline was separate, seemingly unrelated to any other. Before the revolution, paleontologists had no need to talk to oceanographers; the study of volcanoes had little to do with ore geology; geophysicists were unconcerned with life's origins and evolution; and the rocks of one nation were of no obvious relevance to rocks of another, much less to rocks on the distant ocean floor.

Plate tectonics unified everything about Earth. Now locations of rare fossil organisms can be matched precisely across the vast spreading oceans. Extinct volcanic terrains lead miners to valuable ore deposits hidden in their corresponding subduction zones, long since solidified into continental rock. Geophysical studies of shifting continents point to key influences on the evolution of plants and animals. Plate tectonics reveals Earth as an integrated planetary system, from crust to core, at scales from nano to global, with a single unifying principle through space and time.

It took time for the production of granite to shift from the disorderly, vertical tectonic patchwork of plume-driven islands to the coordinated assembly of subduction-driven continents. But by the time Earth was 1.5 billion years old, the convecting mantle—the eighteen-hundred-mile-thick zone that holds most of Earth's mass and heat energy—had irrevocably transformed the surface of our planet. Unlike black basalt, the growing, barren granite landmasses appeared whitish gray, the typical blended color of quartz and feldspar. And so if you were a time traveler to that ancient world three billion years past, you would find some familiar features. You could stand on proto-continents devoid of vegetation, with jagged hills and steep-walled valleys, not unlike some rugged coastlines of the high arctic. You would experience periods of violent weather, punctuated by days of sunny blue skies and puffy white clouds. You would find the ocean saturated with dissolved minerals, including calcium and magnesium carbonates, which were occasionally deposited as crystal layerings on the basalt seafloor. You could lie by that cool, blue ocean on the first white sandy beaches, rich in resistant quartz grains eroded from the gray granite. But you would quickly suffocate in the heavy atmosphere, rich in nitrogen and carbon dioxide but lacking the slightest whiff of life-giving oxygen.

The invention of continents—landmasses constructed of sturdy granite crust—was but a sideshow in Earth's grand evolutionary pageant. Granite terrains, formed by deep heating and partial melting of the ubiquitous near-surface basalt, were like growing gray scabs on the otherwise pristine, submerged black skin of our planet. Gradually, the thickening rafts of granite, floating as they did on top of the denser basaltic basement, were able to rise above ocean level and thus provided the roots of all the great continents—what we perceive today in our anthropocentric view as the solid Earth.

Chapter 6

Living Earth

The Origins of Life

Earth's Age: 500 million to 1 billion years

  Infant Earth at age five hundred million years gave little hint as to how very precocious it would soon become. Earth boasted dramatic volcanism, to be sure, but so did several other planets and moons in our Solar System. Earth was graced by globe-spanning oceans, but so was Mars in those early days, while Jupiter's giant moons Europa and Callisto were enveloped by ice-covered oceans more than fifty miles deep, thus holding far greater proportions of the precious liquid at the surface. Plate tectonics helped to transform our planet, but in those early years, Venus and possibly Mars had their own versions of convection-driven tectonics.

Nor did chemistry set Earth apart. Basalt and granite were the foundation stones for all the rocky planets. Oxygen, silicon, aluminum, magnesium, calcium, and iron dominated their compositions. Earth had its share of carbon and nitrogen and sulfur, but other worlds in our Solar System were equally endowed with those vital elements. By almost every measure, Earth four billion years ago appeared to be a rather ordinary planet.

But Earth was soon to become unique among known worlds. Admittedly, it was already unique in that at five hundred million years, no other known planet or moon had endured such sweeping episodes of change; no other planet had altered its outward appearance so thoroughly and so often. But these metamorphoses were different only in scale, not in kind. The most dynamic engine of planetary change—what sets Earth apart—was yet to emerge. Only Earth became abundantly, persistently alive. The origin and evolution of the biosphere distinguishes Earth from all other known planets and moons.

What Is Life?

What does it mean to be alive? What is this phenomenon that makes Earth so different from the rest of the known cosmos? We might well try to describe life as a set of distinctive intertwined traits—a complex structure, coupled with the ability to move, to grow, to adapt, and to reproduce. We might point to such distinctive cellular attributes as a membrane or long strands of the genetic molecule DNA. But no matter how long the list of diagnostic traits, there always seem to be exceptions. Lichens don't move. Mules don't reproduce.

Chemistry provides a firmer foundation for defining life, for all living things are organized molecular systems that undergo chemical reactions of astonishing intricacy and coordination. Every life-form consists of discrete assemblages of molecules (cells) separated by a molecular barrier from the outside (the environment). These clever collections of chemicals have evolved two interdependent modes of self-preservation—metabolism and genetics—that together unambiguously distinguish the living from the nonliving.

Metabolism is the varied suite of chemical reactions that all life-forms use to convert atoms and energy from their surroundings into more cell stuff. Like tiny chemical factories, cells take in molecular raw materials and fuel and use those hard-won resources to facilitate movement, repair, growth, and, from time to time, reproduction. And like chemical factories, as opposed to raging forest fires or the nuclear chain reactions of the first element-generating star, cells exquisitely control and regulate these reactions by positive and negative feedbacks.

Metabolism alone isn't enough to define life. Unlike their nonliving environment, cells carry information in the form of DNA molecules, and they can copy and pass that molecular information from one generation to the next. What's more, the information can mutate; molecules are often copied with errors, which provide genetic variations. Mutations thus promote chemical novelty—innovations that enable the population of cells to compete against other less efficient populations, to survive during times of environmental change, or to expand their foothold into new environmental niches.

Thus metabolism and genetics must together characterize living matter. But surprisingly, biologists have failed to devise a single, universally accepted definition of life. NASA's Exobiology Program, tasked with investigating the origins of life and the possibility of it on other worlds, has perhaps come the closest. A 1994 NASA panel chaired by Gerald Joyce of the Scripps Research Institute agreed on a streamlined sentence: "Life is a self-sustaining chemical system capable of undergoing Darwinian evolution."

Joyce, who is a leader in attempts to make life in the lab (part of a futuristic field called synthetic biology), recently achieved this benchmark—a remarkable breakthrough, to be sure. He devised a test-tube-bound collection of thousands of diverse interacting molecules that is both self-sustaining and evolving. Enclosed in glass, this intricate process results in evolving proportions, albeit exact copies, of the diverse molecules that were already present from the start of the experiment. Joyce realized that a chemical system that simply churns out molecular duplicates ad nauseam, even if the relative proportion of those molecules evolves over time, is little more than a molecular Xerox machine. Natural living systems, by contrast, have the ability to mutate and thus potentially do completely new things—to explore new environmental territory, to survive unexpected environmental changes, to perform new tasks, to outcompete neighbors for resources. So Joyce has revised his definition to include the characteristic of novelty: "Life is a self-sustaining chemical system capable of *incorporating novelty and* undergoing Darwinian evolution." What is perhaps most remarkable about this development is that Jerry Joyce, realizing the subtlety of life, modestly amended NASA's definition rather than staking a claim to the historic, if Frankensteinian, distinction of being the first to create life in the lab.

Raw Materials

How did nonliving planet Earth invent the intertwined traits of metabolism and genetics? Most of us in the origins-of-life business suspect that the emergence of the first cell was an inevitable geochemical process. Earth possessed all the essential raw materials. Oceans, atmosphere, rocks, and minerals were rich in the necessary elements: carbon, oxygen, hydrogen, nitrogen, sulfur, and phosphorus. Energy, too, was abundant: solar radiation and Earth's inner heat provided the most reliable sources, but lightning, radioactivity, meteor impacts, and many other forms of energy might have contributed. (And there are, consequently, at least as many theories of life's origins as there are sources of elements and energy.)

On one point just about everyone agrees: carbon, the most versatile element of the periodic table, played the starring role. No other element has such rich molecular designs or such diverse molecular functions. Carbon atoms possess an unmatched ability to bond to other carbon atoms as well as to myriad other elements—notably hydrogen, oxygen, nitrogen, and sulfur—with up to four bonds at once. Carbon can form long chains of atoms, or interlocked rings, or complex branching arrangements, or almost any other imaginable shape. It thus forms the backbone of proteins and carbohydrates, of fats and oils, of DNA and RNA. Only versatile carbon-based molecules appear to share the twin defining characteristics of life: the ability to replicate and the ability to evolve.

Every morsel of food we eat, every medication we take, every structure of our bodies and the bodies of every other living thing, is loaded with carbon. Carbon-based chemicals are everywhere: in paints, glues, dyes, and plastics, in the fibers of your clothes and the soles of your shoes, in the pages and binding and ink of this book, and in energy-rich fuels from coal and oil to natural gas and gasoline. And, as we'll see in Chapter 11, our growing reliance on carbon-based fuels and other chemicals is implicated in troubling shifts in Earth's near-surface environment—changes that are occurring at a pace perhaps unmatched in millions of years.

Still, carbon cannot have undergone the remarkable progression from geochemistry to biochemistry by itself. All of Earth's great transformative powers—water, heat, lightning, and the chemical energy of rocks—were brought to bear in life's genesis.

Step 1: Bricks and Mortar

No one yet knows exactly how (or when) the ancient transition from a lifeless to a living world took place, but basic principles are emerging from focused research at dozens of laboratories around the world. Biogenesis must have occurred as a sequence of steps, each of which added chemical complexity to the evolving world. First the molecular building blocks had to come into existence. Then those small molecules had to be selected, concentrated, and organized into life's essential structures—membranes, polymers, and other functional components of a cell. At some point, the collection of molecules had to make copies of itself, while devising a means to pass genetic information from one generation to the next. And then evolution by Darwinian natural selection took over; life emerged.

The first and best-understood step in biogenesis was the rampant production of life's molecular building blocks: sugars, amino acids, lipids, and more. These essential chemicals, all based on the versatile element carbon, emerge anywhere that energy interacts with simple molecules like carbon dioxide and water. Life's raw materials formed where lightning pierced the atmosphere, where volcanic heat boiled the deep ocean, even where ultraviolet radiation bathed molecular clouds in deep space before Earth was born. The seas of ancient Earth became increasingly concentrated in the stuff of life, as biomolecules rained from the skies and rose from the depths.

Modern origins-of-life research began in 1953, with what remains to this day the most famous experiment in biogenesis. Chemist Harold Urey, a Nobel Prize–winning professor at the University of Chicago, and his resolute graduate student Stanley Miller designed a simple and elegant tabletop glass apparatus to simulate early Earth. Gently boiling water proxied for the hot Hadean ocean, and a mixture of simple gases mimicked Earth's primitive atmosphere, while electric sparks simulated lightning. After a few days, the confined, colorless water turned pinkish, then brown, with a complex mix of organic molecules. The transparent glass became smeared with sticky black organic sludge.

Miller's routine chemical analyses revealed an abundance of amino acids and other bio-building blocks. His 1953 paper in *Science,* announcing the results, generated sensationalistic headlines around the world. Chemists soon flocked to the study of prebiotic chemistry. And while the exact combination of atmospheric gases in the Miller-Urey experiment was called into question, thousands of subsequent experimental variations on the theme established beyond any doubt that early Earth must have abounded in life's essential molecules. Indeed, the 1953 spark experiment and its progeny were so successful that many in the field thought the origins mystery had been largely solved.

This initial enthusiasm and subsequent focus may have come at a price. Miller's masterful experiment placed origins-of-life research squarely in the camp of organic chemists and established the paradigm of life emerging from a prebiotic soup—perhaps from a "warm little pond" (echoing Charles Darwin's private speculations from almost a century before). Few experimentalists of the 1950s considered the staggering complexities of natural geochemical environments, altered as they are by daily cycles of night/day, hot/cold, wet/dry, and more. Nor did they consider the range of natural gradients—in temperature, for example, as volcanic magma contacts cold ocean water; or in salinity, as a fresh stream enters the salty ocean. And none of Miller's experiments incorporated rocks and minerals, chemically diverse with dozens of major and minor elements, and their reactive energetic crystalline faces. Earth's sunlit surface, they assumed, was where all the action must have occurred.

Miller's influence was strong, and he and his followers dominated the origins-of-life community for more than three decades. A flood of publications ensued, new journals arose, and honors and awards were bestowed, while government funding flowed to the "Millerites." Then in the late 1980s, the discovery of deep-sea black smoker ecosystems gave rise to a viable alternative to "primordial soup." In those deep dark zones, far from the sunlit ocean surface, mineral-rich fluids interact with hot volcanic crust to generate ocean-floor geyserlike vents. Jets of scalding water contact the frigid deep ocean to create a constant precipitation of minerals (the microscopic particles that produce the black "smoke"). Life abounds in those astounding hidden places, fueled by the chemical energy at the interface between crust and ocean.

The battle over origins paradigms reveals a lot about the sociology of science. On the one hand, the Miller-Urey process produced a suite of biomolecules stunningly similar to what life actually uses. The mix of amino acids, carbohydrates, lipids, and bases almost looks like a well-balanced diet. As Harold Urey quipped, "If God did not do it this way, then He missed a good bet." But the true believers of the Miller camp did more than just support the lightning-seeded primordial soup idea; with a vengeance, they publicly rejected any and all competing ideas.

The effectiveness of the La Jolla cabal's obstructionism began to decline with the startling discovery of those living black smokers described above, coupled with the powerful influence and far-reaching ambitions of NASA. The existence of black smokers at undersea vents underscored a growing awareness that life abounds in extreme environments—in places where a previous generation of biologists would not have looked. We now know that microbes thrive in acidic streams flowing from mine waste and boiling pools above volcanic zones. They eke out a living inside frozen Antarctic rocks, and they persist on stratospheric dust particles miles above Earth's surface. Vast microbial ecosystems miles below Earth's solid surface, where cells live in the narrowest of cracks and fissures and subsist on the meager chemical energy of minerals, may well account for half of Earth's biomass—as much as all the trees and elephants and ants and people combined. If such extremophile life can thrive—if a significant fraction of Earth's life survives in deep environments protected from the violent insults of asteroids and comets—why couldn't life have originated there?

NASA, whose science funding is tied closely to the prospect of great discoveries, jumped at this possibility. If life is constrained to arise in a Miller-Urey scenario, at the sun-drenched surface of a watery world, then Earth and possibly Mars (in its earliest stages, its first five hundred million years) are the only plausible living worlds within our reach. But if life can emerge from the black, hot depths of a subsurface volcanic zone, then many additional celestial bodies become tempting targets for exploration. Mars today must have deep hydrothermal zones; perhaps endures lives there even now. Several of Jupiter's moons are also ripe for biological investigation, as is Saturn's organic-rich, Earth-size moon Titan. Even some of the larger asteroids may have deep, life-producing, hot wet zones. If life arose deep on Earth, then NASA's search (and funding) for exobiology will surely last for many decades.

My Carnegie Institution colleagues and I are relative latecomers to the origins game. Our lab's first NASA-sponsored experiments in 1996 were specifically designed to test organic synthesis in black smoker regimes, where high temperatures and pressures prevail. Like Miller, we subjected mixtures of simple gases to energetic conditions—in our case, heat and chemically reactive mineral surfaces, just as you'd find in a deep volcanic zone. Like Miller, we produced amino acids, lipids, and other bio-building blocks. Our results, now duplicated and expanded in numerous labs, show beyond a doubt that a suite of life's molecules can be synthesized easily in the pressure-cooker conditions of the shallow crust. Volcanic gases containing carbon and nitrogen readily react with common rocks and seawater to make virtually all of life's basic building blocks.

What's more, these synthesis processes are governed by relatively gentle chemical reactions called reduction and oxidation reactions, or redox reactions, such as the familiar rusting of iron or toasting a marshmallow. These are the same kinds of chemical reactions that life uses in metabolism, in sharp contrast to the violent ionizing effects of lightning or ultraviolet radiation. Indeed, while harsh lightning bolts may facilitate the production of small biomolecules, they just as easily rip those building blocks to molecular shreds. To many of us in the origins game, it makes a lot more sense for Earth to have made its prebiotic molecules with less energetic chemical reactions, in more or less the same way that cells do it today.

Stanley Miller and his followers did what they could to squelch our conclusions and abort our research program. In a flurry of critical publications, they argued that the high temperatures of the volcanic vents would quickly destroy any useful biomolecules. "The vent hypothesis is a real loser," Miller complained in a 1998 interview. "I don't understand why we even have to discuss it." They based their arguments on meticulous experiments in which biomolecules degrade in boiling water. But these simplistic studies failed to mimic the complexity of primordial Earth; missing were the deep ocean's extreme gradients of temperature and composition, the turbulent flow and cycling of volcanic vents, the chemical complexity of mineral-rich seawater, or the protective surfaces of rocks on which biomolecules are now known to bind. Nonetheless, the origins field has now moved beyond the Miller-Urey scenario, and for many experts, Earth's deep dark zones are today the primary focus in the biogenesis game.

As I mentioned earlier, *every* ancient environment with sources of energy and small carbon-bearing molecules probably produced its share of amino acids, sugars, lipids, and other molecular building blocks of life. An atmosphere laced with lightning or exposed to harsh radiation remains in the running as a theory of biogenesis; so do black smokers and other deep, hot environments. Biomolecules form during asteroid impacts, on sun-drenched dust particles high in the atmosphere, and in deep-space molecular clouds exposed to cosmic rays. Every year tons of organic-rich dust rain down on Earth's surface from outer space, as it has for more than 4.5 billion years. We now know that life's building blocks litter the cosmos.

Step 2: Selection

A half-century ago, the greatest challenge in origins research was synthesizing the raw materials: the molecular bricks and mortar of life. By the dawn of the twenty-first century, that problem had largely been solved; scientists realized that Earth must have been girdled by a dilute consommé of life's vital ingredients. Much of the focus has now shifted to the selection, concentration, and assembly of biobits into macromolecules—to the membranes that enclose the cell, the enzymes that promote its chemical reactions, and the genetic polymers that pass information from one generation to the next.

Two complementary processes likely played a role. One is self-assembly, in which a group of elongated molecules—lipids—clumped together spontaneously to form the membranes that encapsulated the first cells. Lipids feature skinny backbones of a dozen or more carbon atoms. Under certain conditions, they tend to self-assemble into microscopic hollow balls; the elongated molecules line up side to side, like seeds on a dandelion head. In one of the most influential origins publications in history, California biochemist David Deamer described how he extracted a suite of these versatile organic molecules from the carbon-rich Murchison meteorite (a conglomeration of chemicals formed in deep space long before Earth) and found that they rapidly organized themselves into tiny cell-like spheres with an inside and an outside, not unlike tiny oil drops in water. A few years ago Deamer and I found that carbon-rich molecules forming under hot, pressurized black smoker conditions behave in much the same way. These and other experiments reveal that membrane-bound vesicles are an inevitable feature of the prebiotic world; lipid self-assembly must have played a key role in life's origins.

Most other bio-building blocks don't self-organize, but they can become concentrated and arrayed on the safe, protective surfaces of rocks and minerals in what's known as template-directed synthesis, the second of the two selection processes. Our experiments, conducted at the Carnegie Institution over the past decade, reveal that many of life's most vital molecular building blocks stick to virtually any natural mineral surface. Amino acids, sugars, and the components of DNA and RNA adsorb onto all of Earth's most common rock-forming minerals in basalt and granite: feldspar, pyroxene, quartz, and others. What's more, when several molecules compete for the same piece of crystal real estate, they often cooperate and yield complex surface structures of their own that may promote even more adsorption and more organization. We concluded that wherever the prebiotic ocean contacted minerals, highly concentrated arrangements of life's molecules are likely to have emerged from the formless broth.

Here I should issue a caveat. In origins-of-life research (and probably in most other disciplines as well), scientists gravitate to models that highlight their personal scientific specialty. Organic chemist Stanley Miller and his cohorts saw life's origins as essentially a problem in organic chemistry. Geochemists, by contrast, have tended to focus on more intricate origins scenarios involving such variables as temperature and pressure and chemically complex rocks. Experts in membrane-forming lipid molecules promote the "lipid world," while molecular biologists who study DNA and RNA view the "RNA world" as the model to beat. Specialists who study viruses, or metabolism, or clays, or the deep biosphere have their idiosyncratic prejudices as well. We all do it; we all focus on what we know best, and we see the world through that lens.

I'm trained in mineralogy, so you can easily guess where my origins preferences lie. Mea culpa. Many other origins researchers have also settled on such a conclusion—indeed, more than a few prominent biologists have also gravitated to minerals, because origins-of-life scenarios that involve only oceans and atmosphere face insurmountable problems in accounting for efficient mechanisms of molecular selection and concentration. Solid minerals have an unmatched potential to select, concentrate, and organize molecules. So minerals must have played a central role in life's origins.

Right and Left

Biochemistry is complex, with interwoven cycles and networks of molecular reactions. For these intricately layered processes to work, molecules have to have just the right sizes and shapes. Molecular selection is the task of finding the best molecule for each biochemical job, and template-directed selection on mineral surfaces is now the leading candidate for how nature did it.

Perhaps the most daunting challenge in molecular selection is chirality, the pervasive "handedness" of life. Many of life's molecules come in mirror-image pairs—left- and right-handed variants like your two hands. Chiral pairs of molecules are identical in many respects: they have the same chemical compositions, the same melting and boiling points, the same color and density, and the same electrical conductivity. But left- and right-handed molecules have different, incompatible shapes—a familiar characteristic if you've ever tried to put a left-handed glove on your right hand. It turns out that life is incredibly picky: cells almost exclusively employ left-handed amino acids and right-handed sugars.

Chirality matters. In the curious case of the artificial fragrance limonene, the right-handed form smells like an orange, whereas the left-handed version of this simple ring-shaped molecule smells like a lemon. The smell receptors in your nose are sensitive to chirality, so right- and left-limonene transmit slightly different signals to your brain. Taste buds are less sensitive to the differences between right- and left-handed sugars. They both taste sweet, but our body's fine-tuned digestive system can process only the right-handed forms. The artificial sweetener tagatose, a zero-calorie left-handed sugar substitute, exploits these properties. The tragic story of thalidomide also rests on handedness. The right-handed version of this drug alleviated morning sickness in pregnant women, but the left-handed variant that inevitably tagged along caused birth defects. Today the FDA imposes strict requirements for chirally pure drugs—regulations that save lives but cost consumers an estimated $200 billion per year in added manufacturing costs.

Most experiments that synthesize biomolecules (including Miller-Urey and hydrothermal experiments) produce equal amounts of left- and right-handed molecules, and most natural processes treat left- and right-handed molecules exactly the same. Indeed, the nonliving natural world is for the most part indifferent to the distinction between left and right. But life absolutely requires the correct shape: left-handed amino acids and right-handed sugars are essential. The opposite-handed molecules simply will not do. So our research team tackled the question of how life selected left-handed amino acids almost exclusively over right, and right-handed sugars over left.

Our recent experiments have explored the possibility that chiral mineral surfaces played the starring role in selecting handed molecules, and perhaps in the origins of life as well. In 2000 my colleagues and I realized what was then surprising but is now obvious: chiral mineral surfaces are everywhere in nature. The commonest minerals in every rock and every soil abound with surfaces where atoms form molecular-scale "handholds," some left and some right. In the natural world, these left- and right-handed mineral surfaces occur in statistically equal proportions, so Earth on a global scale doesn't appear to be biased for either left or right. But each individual molecule does care where it winds up. Our experiments showed that certain left-handed molecules can aggregate on one set of crystal surfaces, while the mirror-image, right-handed counterpart molecules aggregate just as easily on other sets of mineral surfaces. As handed molecules are separated and concentrated, each surface becomes a tiny experiment in molecular selection and organization.

On its own, no such natural experiment with minerals and molecules is likely to have generated life. But take countless trillions of trillions of trillions of mineral surfaces, each bathed in molecule-rich organic broth, and repeat those tiny natural experiments over and over for hundreds of millions of years. Earth must have eventually tested virtually every combination of small molecules somewhere, sometime. The tiny fraction of all those molecular combinations that wound up displaying easier self-assembly, or developed a stronger binding to mineral surfaces, or enjoyed greater stability under the high temperatures and pressures, survived, possibly to grow, possibly to learn new tricks.

We don't yet know exactly which of those myriad possible combinations of molecules and minerals led to lifelike organization, but the principles of molecular selection and organization are now emerging. Biomolecules were synthesized in abundance, and some of those molecules went on to form larger and larger clusters. Our experiments suggest that electric charge played a big role. Some molecules have a slight positive charge; others have a slight negative charge; and still others (like water) are polar with slightly positive and negative ends to the same molecule. Minerals, too, have charged surfaces, some positive and some negative. Add all these charged pieces together, and they spontaneously organize, with positive electric charge always attracted to negative electric charge. And so varied molecular assemblies occurred in virtually every wet, mineral-rich environment on prebiotic Earth.

Step 3: Replication

Arrays of chemicals, no matter how intricately patterned, are not alive unless they can make copies of themselves. Life's most distinctive hallmark is reproduction: one consortium of molecules becomes two, two become four, and on and on in geometric expansion. The greatest enigma in the biogenesis story remains the emergence of that first system of self-replicating molecules. Clever experiments replicate portions of plausible reproductive cycles, though we have yet to mimic completely that elusive biochemical trick in the lab. Nevertheless, at some point in space and time, an organized collection of molecules began to duplicate itself at the expense of other molecules (that is, "food").

Imagine Earth at age five hundred million years—roughly four billion years ago. It had a broth of organic molecules, it had trillions upon trillions of reactive mineral surfaces, and it had hundreds of millions of years to play with. Most of the molecular milieu did nothing of interest and displayed no useful functions. But a small fraction of the organic molecules arrayed on mineral surfaces produced some kind of structure with enhanced function—perhaps a stronger surface attachment, or maybe the means to attract more molecules to the community, or the tendency to catalyze the destruction of competing molecular species, or even the ability to make copies of itself. The natural world amply rewards such innovation, and once established, life quickly infested every habitable nook and cranny of the globe.

But let's take a step back. Why would a collection of molecules spontaneously start copying itself? The answer lies in the twin evolutionary pillars of variation and selection. Systems evolve for two reasons. First, they display vast numbers of different possible configurations—that's variation. Second, some of those configurations are much more likely to survive than others—that's selection. Imagine a prebiotic collection of hundreds of thousands of different molecules, all made of carbon, hydrogen, oxygen, and nitrogen, maybe with some sulfur or phosphorus thrown in. Prebiotic synthesis (à la Stanley Miller) and natural samples (for example, David Deamer's meteorite) display this degree of molecular variation. But not all molecules were created equal. Some molecules were relatively unstable and decomposed—they were quickly eliminated from the competition. Others clumped together in useless tarlike masses and floated away or sank to the ocean floor, where they could play no further role. But some molecules proved to be especially stable, perhaps even more so when they could bind to others of their kind or to a particularly tempting mineral surface. These molecules survived, as the molecular broth was relieved of the least fit.

Molecular interactions further refined the prebiotic mix. Some groups of molecules cooperatively stuck to mineral surfaces, thus enhancing survival of the clique. Other small molecules acted as catalysts, enhancing some chemical species by promoting formation of chemical bonds, or speeding the destruction of other chemical species by breaking chemical bonds. The molecular broth was swiftly winnowed, but ultimate security in such a world was not to be found in eliminating the competition or just hanging on. The ultimate prize of survival would go to the collection of molecules that learned to make copies of itself.

Three competing models attempt to describe the first self-replicating, quasi-living system of molecules. The simplest of these models (and therefore the one many of us prefer) points to a well-known cycle of a few small molecules—the ubiquitous citric acid cycle. It starts with acetic acid, which contains only two carbon atoms. Acetic acid reacts with CO_2 to form pyruvic acid (with three carbon atoms), which in turn reacts with more CO_2 to make the four-carbon oxaloacetic acid. Other reactions produce progressively larger molecules, up to citric acid, with its six carbon atoms. The cycle becomes self-replicating when citric acid spontaneously splits into two smaller molecules, acetic acid (two carbon atoms) plus oxaloacetic acid (four carbon atoms), which are also part of the molecular loop. One cycle of molecules thus becomes two, two become four, and so on. What's more, many of life's essential building blocks, including amino acids and sugars, are readily synthesized by simple reactions with the core molecules of the citric acid cycle. Just add ammonia to pyruvic acid, for example, and you get the essential amino acid alanine. Every living cell on Earth incorporates the citric acid cycle, so it may well be a primordial characteristic—a chemical fossil descended from the very first life-form. Such a cycle, by itself, is not alive. But it does have the potential to replicate the inner circle of molecules at the expense of less fecund chemicals.

At the opposite extreme of chemical complexity is the self-replicating autocatalytic network, a model championed by Stuart Kauffman, who conducted pioneering theoretical studies at the famed Santa Fe Institute. The prebiotic broth may have initially incorporated hundreds of thousands of different kinds of small, carbon-based molecules from varied sources. We now know that some of those chemicals catalyzed reactions that made new molecules, while other reactions accelerated the breakdown of their neighbors. An autocatalytic network consists of a collection of molecules—perhaps thousands of different species working in concert—that speed up the production of themselves, while destroying any molecule not in the network. It's the molecular equivalent of "the rich get richer." Again, as with the citric acid cycle, such a molecular network would not be considered to be alive, but in a way it does promote the copying of itself, and it is far more complex than most nonliving chemical systems.

A third scenario, the one probably favored by the majority of biologically trained origins researchers, is the RNA world—a model based on a hypothetical molecule of RNA that makes copies of itself. To understand why this scenario appeals, we need to take another step back, to think about life's two most critical functions: metabolism (making stuff) and genetics (transferring information on how to make stuff from one generation to the next). Modern cells use the ladderlike molecule DNA to store and copy the information needed to make more proteins, but they use complexly folded protein molecules to make the DNA. So which came first, DNA or proteins? It turns out that a third kind of molecule, RNA, plays the central role in both processes.

RNA is an elegant polymer—a long, single-stranded molecule assembled from smaller individual molecules (called nucleotides) like beads on a string or letters in a sentence. Four different molecular "letters," designated A, C, G, and U, can line up in any imaginable sequence, like a coded message. Indeed, these RNA letters hold genetic information (just like DNA). At the same time, RNA can fold up into complex shapes that have the ability to catalyze key biological reactions (just like proteins). In fact, RNA molecules facilitate the synthesis of all proteins, both by carrying genetic information and by catalyzing protein formation. So of all life's varied molecules, RNA is the only one that seems to "do it all."

The RNA world model rests on the assumption that some as yet poorly understood chemical mechanism produced vast numbers of different strands of RNA, or perhaps an information-rich molecule very similar to it. Almost all those varied strands did absolutely nothing; they simply survived or gradually degraded. However, a select few strands possessed some kind of self-beneficial function; they folded up to become more stable, or they latched on to a secure mineral surface, or perhaps they destroyed their rivals—just another example of molecular competition in the broth.

The key assumption of the RNA world hypothesis is that one of these myriad strands learned the remarkable trick of how to make copies of itself: it became a self-replicating molecule. This idea isn't so far-fetched. After all, RNA is a lot like DNA, which is able to make copies of itself. What's more, RNA is easily mutated. So the first self-replicating RNA molecule, however inefficient or sloppy, would soon have found itself competing with lots of slight variants of itself, some of which pulled off the copying trick a little faster, or with a little less energy expenditure, or perhaps in slightly different environments. Such a precocious RNA molecule would seem to fulfill all the requirements for life: it is a self-sustaining chemical system capable of incorporating novelty and undergoing Darwinian evolution—in this case, molecular evolution.

Perhaps it took a long time for that first functional, crudely self-replicating molecular system to emerge, whether it was a citric acid cycle, or an autocatalytic network, or self-replicating RNA. But unimaginable numbers of molecular combinations were being tried on trillions of trillions of mineral surfaces, across almost two hundred million square miles of Earth's surface, for many millions of years. And one of those inconceivably immense numbers of molecular combinations, someplace, sometime, worked. It learned to self-replicate and to evolve. And that invention changed everything.

Experiments in the Boston-based laboratory of Harvard biologist Jack Szostak demonstrate the power of selection in molecular evolution. In many of their experiments, Szostak's team starts with a mixture of one hundred trillion different RNA sequences, each of which consists of a random hundred-letter string of A, C, G, and U. That immense collection of diverse RNA strands, each of which folds up differently, is then confronted with a task: for example, bonding tightly to another distinctively shaped molecule. Szostak's team pours a solution with all hundred trillion strands into a beaker with little glass beads, each of which has been coated with that distinctively shaped target molecule. These target molecules dangle out into the RNA-rich solution like little hooks. The vast majority of RNA molecules don't respond; they have the wrong shapes to interact. But a tiny fraction of folded RNAs latch on to the targets and stick tight.

That's where the fun begins, because Szostak's coworkers pour out the old solution (along with almost one hundred trillion different nonfunctional RNA strands) and recover the few strands that by virtue of their fortuitous shapes happen to stick to the coated glass beads. Then, using standard tricks of genetic technology that mimic plausible prebiotic processes, they prepare a new batch of one hundred trillion RNA strands, but this time all the strands are sloppy copies—each a mutant of one of those few original functional strands. Repeating the above steps produces a new population of functional RNA strands, but some of these second-generation variants bind much better than any of the first generation. Some of the mutant daughter strands significantly outperform their parents. Repeat the whole process a few more times, and the resultant RNA strands get better and better at binding, until the best mutants are perfectly functional: they lock on to their targets with the highest possible binding energy.

The whole experiment takes a few days—less than a week from random strands to a perfect binding molecule. But if you were to ask a team of the world's most brilliant chemists to design such a functional RNA strand from scratch, they would find the task to be virtually impossible with any known computational method. No current method can predict exactly how a long RNA strand will fold up, or how it might attach to other complexly shaped molecules. Molecular evolution, not intelligent design, is by far the fastest and most reliable path to achieving function. (That's why we say if God created life, she's smart enough to use evolution.)

Life's Explosion

In the prebiotic broth, a collection of molecules with even the slightest useful function had an advantage. But these molecular games of war paled beside the advantage conferred upon the RNA strand that had a useful function *and* could make copies of itself. Such a self-replicating molecule ensured its own survival by producing more or less identical daughters. What's more, the molecular copying process was inevitably messy, so some of those RNA copies were mutants. And while most mutations were lethal or conferred no significant advantage, a few fortuitous individuals outshone their parents, and thus the system evolved. Simply by chance copying errors, the original self-replicating molecule must have produced offspring that tolerated more extreme conditions of pressure or heat or salt, or replicated faster, or found new sources of food, or destroyed their less fit neighbors. Even greater advantages were enjoyed by those RNA strands that found protection on a mineral surface or refuge within a safe, encapsulating membrane.

Without competition, the first self-replicating molecules engulfed Earth's nutrient-rich zones in a geological instant. It's perhaps counterintuitive to think of a microscopic object taking over, but say, just for argument, that the first relatively inefficient self-replicating molecule took a week to duplicate once. (Many modern microbes, by contrast, can replicate in a matter of minutes.) Week by week, two strands became four, four became eight, and so forth. At that rate, it would have taken about half a year to form a clump of one hundred million self-replicating molecules—an object just large enough to see with the naked eye. In another twenty weeks, the mass of RNA would have expanded to fill a thimble. And at this rate, it would take yet another twenty weeks for all of life's earliest manifestation to fill a good-size bathtub.

But continued doubling every week would have quickly produced a remarkable transformation. Twenty more weeks would have seen miles of RNA-infested waters, perhaps along the coast or in an inland lake or a deep-sea environment. And within two years, assuming that a single initial RNA strand doubled every week, Earth could boast as much as a million cubic kilometers of living stuff—enough to clog the entire Mediterranean Sea.

Primitive single-celled organisms that fed on the chemical energy of rocks couldn't have had much effect on Earth's geology—its near-surface distribution of rocks, for example, or the diversity of minerals. Living or no, four billion years ago the ancient land remained barren black and gray, surface weathering was slow, and the earliest life would have contributed almost nothing to alter the globe-spanning blue oceans.

Because the first scrappy microbes would have made little mark, we can't tell for sure when life began. A few of Earth's most ancient sedimentary rocks, those laid down in shallow ocean environments about 3.5 billion years ago, hold unmistakable microbial fossils. Domelike rocky stromatolites, a few inches to a few feet across, formed where colonies of cells precipitated thin layer upon layer of minerals in shallow settings. Microbial mats covered wide swaths of shoreline, consolidating and patterning sand in tidal zones. Even a few carbon-rich spheres with distinctive cell-like walls—possible microbial body fossils—have survived the aeons. But no older incontrovertible fossils have been found. Geochemical traces of carbon and other bioelements from intensely altered rocks 3.85 billion years old are tantalizing, but by no means have they convinced the geological community.

So when did life arise? If your hunch is that life emerges early and often on any viable planet or moon, then perhaps you'd advocate a stable biosphere by 4.4 billion years ago—within Earth's first 150 million years. All the ingredients were there: oceans and air, minerals and energy. Giant impacts of asteroids and comets would have challenged the survival of this Hadean life, perhaps favoring those hardy cells that learned to live hot and deep in protected rocky homes beneath the ocean floor. Perhaps life came into being more than once, perhaps many times, before Earth settled into its calmer postadolescence. If so, then the 3.5-billion-year-old fossils would represent an ecosystem almost a billion years in the making.

If, on the other hand, you suspect that life's origins are difficult and rare in the cosmos, then a date closer to 3.5 billion years ago may seem more likely. Perhaps life is so improbable that a billion years of mineral-molecule interactions throughout hundreds of millions of cubic miles of ocean crust were necessary. Perhaps those precious, sparse fossil remains of the so-called Archean Eon mark the true beginnings of the biosphere.

Living Earth

Whenever first life emerged, whether before 4.4 billion years or after 3.8 billion years ago, the fact remains: it little altered Earth's ancient surface. Those earliest microbes simply learned chemical tricks that Earth already knew. From our planet's earliest days, chemical reactions have taken place at or near its solid surface. The reason boils down to the distribution of electrons: Atoms in Earth's mantle have on average more electrons poised to engage in chemical reactions than atoms in the crust. The mantle is more "reduced" and the surface more "oxidized," in the jargon of chemistry. When reduced and oxidized chemicals meet—for example, when reduced magma and gases from the mantle breach the more oxidized surface in a volcanic eruption—they often undergo an energy-liberating chemical reaction. In the process, electrons transfer from the former to the latter.

Rusting, in which iron reacts with oxygen, is a familiar example of such a reaction. Iron metal is packed full of electrons—so many electrons, you'll recall, that some of them are free to wander through the shiny metal and conduct electricity. Iron is thus an electron donor. Oxygen gas, on the other hand, is so starved for electrons that pairs of oxygen atoms must pool their resources to make an O_2 molecule, in which the meager supply of electrons is shared by both atoms like rations on a desert island. Oxygen is the ideal electron acceptor. So when iron metal meets molecules of oxygen, a rapid exchange of electrons occurs. Each iron atom gives up two or three electrons, while each oxygen atom takes up two electrons. The result of this exchange is a new chemical compound, iron oxide, plus a little jolt of energy.

In addition to iron, the common electron-sated metal elements nickel, manganese, and copper were subject to oxidation. So, too, were many of the simple carbon-based molecules that had been synthesized in prebiotic processes, including methane (natural gas), propane, and butane. Oxygen gas was scarce in Earth's earliest atmosphere, but other electron-hungry collections of atoms, including sulfate (SO_4), nitrate (NO_3), carbonate (CO_3), and phosphate (PO_4), were readily available to fill that role.

Before life's emergence, redox reactions proceeded at a relatively leisurely pace. But the first microbes learned to shuffle electrons at an accelerated rate. In many places—primitive coastlines, near-surface waters, ocean-floor sediments—living cells became the mediators of these reactions. Communities of microbes made their livelihoods by speeding up reaction rates of the rocks, using the resultant energy to live and grow and reproduce. Earth had made iron oxides from the start, to be sure, but the first microbes made them faster. In the process, life began, ever so slowly, to alter Earth's surface environment. Microbes exploited the abundant energy available, in the form of the reduced iron dissolved in the Hadean and Archean oceans; they oxidized iron to form the rusty red mineral hematite—a chemical transformation that can release enough energy to support an entire ecosystem. Massive Archean banded iron formations found in Australia, South Africa, and other ancient terrains may thus represent the leavings of an epic microbial buffet that lasted tens of millions of years. And so began the astonishing coevolution of the geosphere and biosphere.

Evolution by natural selection continued to drive all these processes. A microbial species that learned to use its iron food source more efficiently, or to tolerate more extreme conditions, or to exploit new redox reactions, had a distinct advantage and ensured its own survival. Consequently, mutating populations of microbes invented new catalysts that promoted these energy-producing reactions more efficiently than the nonliving environment. The results, here and there, were little mounds of limestone and modest deposits of iron oxides, along with a gradual increase in the processing of near-surface carbon, sulfur, nitrogen, and phosphorus. Nevertheless, these earliest life-forms did little more than mimic the chemistry that had already begun (albeit more slowly) on the previously nonliving world.

Light

Most origins researchers suspect that the earliest life-forms relied exclusively on the chemical energy of rocks—an abundant source, to be sure, but one that greatly restricted where life could thrive. At some point, a few microbes moved beyond their role as mediators of chemical reactions intrinsic to their environment. They learned to collect solar radiation, which would prove an abundant, cheap energy source for any surface dweller, anywhere on the planet.

In its most basic form, photosynthesis uses sunlight to make biomolecules from such ubiquitous raw materials as carbon dioxide, nitrogen, and water. Given the right chemical scaffolding, all of life's essential building blocks—amino acids, sugars, lipids, and the components of DNA and RNA—can be made from atmospheric gases and the Sun's radiation. Unlike modern green algae, those first microbial photosynthesizers generated no oxygen. Indeed, modern analogues of these primitive organisms tend to form a brownish or purplish scum on stagnant ponds. Some biologists have even suggested that immense floating rafts of photosynthetic microbes discolored the blue Archean oceans with unsightly brownish-purple splotches.

How would we know? Such microbes have no hard parts to preserve as fossils; nor do floating algal mats alter the rock record in any obvious way. However, there may be a way to tease out evidence of the most ancient light-loving microbes. Photosynthetic cells called cyanobacteria rely in part on hopanoids—distinctive molecules with five interlocking carbon rings (a configuration closely related to that of the steroids so much in the sporting news these days). After microbes die and decay, their telltale multiring hopanoids backbones can survive for billions of years as a molecular residue in fine-grained ocean sediments. Meticulous chemical processing is required to extract and analyze these remnants of hopanoids from the bulk rock. Tentative interpretations must incorporate a litany of tricky assumptions about possible sources of contamination, both ancient and recent. The paleontological community greets every report of molecules surviving for several billions of years with caution, if not outright skepticism. Nevertheless, the chemical traces are there and may be the best window onto this tenuous ancient biosphere. (More on how in chapter 7.)

By our planet's one-billionth birthday, life had established a firm, if relatively inconsequential, foothold on its surface. For another billion years, Earth's microbial life would gently nudge near-surface environments, first by speeding along redox reactions and then through photosynthesis. As near as we can tell, even at two billion years old, Earth would not have displayed any significant life-imposed mineralogical novelty at or near its surface. Cells would simply make more iron oxides, more limestone, more sulfates and phosphates than might otherwise have formed. They would build layered deposits of iron-rich minerals in the deeper ocean and craft protective rocky mounds in coastal shallows—all phenomena that had transpired on Earth before the dawn of life, and on other planets and moons in the Solar System.

But Earth and its primitive microbial population were poised to make the most dramatic transformation in the planet's history. Over the next 1.5 billion years, photosynthesizing microbes would learn a new chemical trick—to exhale the highly reactive, dangerously corrosive gas called oxygen.

EARTH'S AGE *(billions of years)*

0	1	2	3	4	4.567
Hadean Eon	*Archean Eon*		*Proterozoic Eon*	*Phanerozoic Eon*	

⋏

Chapter 7

Red Earth

Photosynthesis and the Great Oxidation Event

Earth's Age: 1.0 to 2.7 billion years

ᖇ Fast-forward to the present day, and it's clear that life has irrevocably transformed Earth's near-surface environment—most conspicuously the oceans and atmosphere but the rocks and minerals as well. It would take more than a billion years, after the innovation of the first living cell, for such a transformation to begin. In that period, new varieties of microbes may have created a brownish or purplish scum in some coastal regions. There may even have been patches of greenish slime decorating equatorial shores and populating shallow ponds, as a few clever cells experimented with new ways to harness the Sun's radiant energy. But the continents were still barren: no plants clung to the rocky landscape, nor were any animals present to eat them. You still would have died quickly, in agony, if you were stranded in this anoxic world.

Earth's surface changed from dull gray to brick red, in a geological afternoon, with the innovation of oxygen-producing photosynthesis and the consequent rise of an oxidizing atmosphere. It's difficult to document exactly when and how quickly slimy green algae evolved to trigger this transformation, called the Great Oxidation Event. Our best guess comes from subtle changes in the rock record, which suggest a pulse of photosynthesis shortly after Earth's two-billionth birthday—about 2.5 billion years ago. After that modest start, things happened relatively quickly: by 2.2 billion years ago, atmospheric oxygen had risen from zero to more than 1 percent of its modern level, forever changing Earth's surface.

The intriguing story of Earth's initial oxygenation is only now coming into focus, as unexpected new clues have emerged and promising new lines of evidence have been pursued in earnest. The past half century of paleo-atmosphere research has seen many competing, sometimes diametrically opposed ideas, but the scientific method is a great winnower of the untenable and the false. We don't yet have the whole narrative, but we are getting much closer, and the picture that is emerging is (literally) breathtaking.

Testimony of the Rocks

Evidence for the Great Oxidation Event comes from a growing catalog of observations on rocks and minerals that date from a vast chunk of Earth's history—roughly 3.5 to 2.0 billion years ago. On the one hand, many rocks older than 2.5 billion years contain minerals that are easily destroyed by the corrosive effects of oxygen, suggesting an oxygen-free environment prior to that time. Geologists find unweathered and rounded pebbles of pyrite (the iron sulfide also known as fool's gold) and uraninite (the commonest uranium mineral) in ancient streambeds—places where such minerals would quickly corrode and break down in today's oxygen-rich surface environment. Such ancient sandy layers also have a telltale chemistry—unusually concentrated in oxygen-avoiding elements like cerium, while strikingly deficient in others like iron compared with today's soils. These chemical quirks further evince an atmosphere devoid of oxygen.

By contrast, rocks younger than 2.5 billion years contain many unambiguous signs of oxygen. Between 2.5 and 1.8 billion years ago, massive deposits of iron oxides called banded iron formations were deposited in staggering abundance. These distinctive, dense accumulations of alternating black and rusty red layers hold 90 percent of the world's known iron ore reserves. Manganese oxides also suddenly appear, as thickly stratified deposits that now provide the world's chief repositories of manganese ores. Hundreds of other new minerals— oxidized ores of copper, nickel, uranium, and more—also appear in the rock record for the first time after the Great Oxidation Event. Yet in spite of this expanded mineralogical repertoire, some scientists remained unconvinced that the Great Oxidation Event was really an event at all. Perhaps there was just a slow and steady increase in atmospheric oxygen. Perhaps the spotty, eroded rock record is incomplete and misleading.

The smoking gun of the Great Oxidation Event came from an unexpected source—remarkable recent data on the isotopes of the common element sulfur. The 1990s saw dramatic increases in the resolving power and sensitivity of mass spectrometers, which are the workhorse analytical instruments of the isotope world. The new generation of mass spectrometers allowed scientists to analyze smaller and smaller samples, even microscopic mineral grains or individual living cells, with higher and higher precision. Sulfur, one of life's essential elements, proved a particularly tempting target for study, as there are four stable sulfur isotopes in nature: sulfur-32, -33, -34, and -36. All of these isotopes have sulfur's requisite sixteen protons in the nucleus, but the number of neutrons varies from sixteen to twenty.

The distribution of sulfur isotopes is usually predictable on the simple basis of mass. All atoms wiggle, but less massive isotopes wiggle more. Consequently, in any chemical reaction light isotopes are more likely to jump about than heavy ones. This selective process, called isotope fractionation, occurs anytime a collection of sulfur atoms experiences a chemical reaction, whether in a solid rock or a living cell. In the case of sulfur, an isotope of mass 32 will typically fractionate more than an isotope of mass 34 or 36. What's more, the fractionation ratio is usually directly related to the mass ratio of the isotopes: the fractionation of sulfur-36 to sulfur-32 is almost always twice the fractionation of sulfur-34 to sulfur-32. This basic physics follows directly from Newton's laws: force equals mass times acceleration. Smaller mass means a bigger acceleration, so under a given force, sulfur-32 wiggles more than sulfur-34, which wiggles more than sulfur-36.

A decade ago geochemist James Farquhar, working at the scenic, oceanside San Diego campus of the University of California, found a profound and unexpected change in the distributions of sulfur isotopes in rocks older than about 2.4 billion years. More recent rocks and minerals almost always display the mass-dependent trend one would expect; their ratios of sulfur isotopes depend almost exclusively on the ratios of mass. But Farquhar and his colleagues saw a radically different fractionation of sulfur isotopes in many rocks older than 2.4 billion years—in some samples, wild deviations of several tenths of a percent (which is a lot in this case). What could have caused such a mass-*independent* deviation from the unassailable Newtonian laws of motion?

Clever theorists, backed by experimental proofs, were quick to point to a solution in the nuances of quantum mechanics. Under the influence of ultraviolet radiation, isotope behavior can deviate from the Newtonian ideal. It turns out that isotopes with an odd mass number, such as sulfur-33, can be selectively affected by ultraviolet (UV) radiation. If a molecule of sulfur dioxide or hydrogen sulfide happens to incorporate sulfur-33, and if that molecule encounters an ultraviolet ray (most likely high in the atmosphere), it may react more readily. The sulfur-33 experiences a "mass-independent fractionation" that skews the isotope ratios.

But why the sudden change on Earth 2.4 billion years ago? The answer lies in the UV-absorbing properties of ozone, a molecule of three oxygen atoms that has been much in the news these past two decades. Today ozone high in the atmosphere provides an essential barrier to the Sun's potentially lethal ultraviolet rays. Measurements taken over the past two decades reveal that this high-altitude ozone layer has been significantly depleted, most likely by destructive reactions with human-produced chemicals called chlorofluorocarbons, or CFCs. (Freon, once used in air conditioners, is the best-known example.) Such an "ozone hole" allows more cancer-causing UV radiation to reach Earth's surface. The good news is that a worldwide ban on CFC production seems to be resulting in a rapid restoration of the ozone layer.

Before the rise of oxygen gas and the consequent cosmic sun-blocking ability of the ozone layer, sulfur compounds high in the atmosphere were subjected to a nonstop bath of ultraviolet radiation. Under such harsh conditions, compounds with sulfur-33 experienced mass-independent fractionation. After the Great Oxidation Event, oxygen in the upper atmosphere reacted with these some of these sulfur compounds and all but wiped out this odd isotope effect.

As lab after lab around the world verified and amplified Farquhar's findings, the great majority of Earth scientists accepted the reality of the Great Oxidation Event. Unless scientists discover a way other than ozone to block ultraviolet rays, the sulfur isotope data peg the beginning of the Great Oxidation Event at about 2.4 billion years ago.

Making Oxygen

So where did all the oxygen come from? These days one of the first topics in any introductory biology class is photosynthesis—the remarkable ability of plants to combine water, carbon dioxide, and sunlight to make their tissues, while producing oxygen as a by-product. We now take it for granted that plants play this central role in making our world a habitable place, but the discovery of photosynthesis was one of the greatest advances in science. And like so many of science's pivotal discoveries, it came in piecemeal fashion.

The discovery of water's role came first. Detailed mechanisms of plant growth were a mystery to scientists of the seventeenth century, but a common assumption held that a plant's tissues must come from the mineral-rich soil, which must therefore be consumed as plants grow. Flemish physician Jan Baptista Van Helmont (1579–1644) put this assumption to the test by conducting a simple experiment in the 1640s. In his own words:

> I took an Earthen Vessel, in which I put 200 pounds of Earth that had been dried in a Furnace, which I moystened with Rain-water, and I implanted therein the Trunk or Stem of a Willow Tree, weighing five pounds; and at length, five years being finished, the Tree sprung from thence, did weigh 169 pounds, and about three ounces: But I moystened the Earthen Vessel with Rain water or distilled water (alwayes when there was need)... At length, I again dried the Earth of the Vessel, and there were found the same 200 pounds, wanting about two ounces. Therefore 164 pounds of Wood, Barks, and Roots, arose out of water onely.

Van Helmont's discovery was a major advance, though water (as we now know) was only part of the story.

A century later the English clergyman and naturalist Stephen Hales first suggested that plants rely on some component of the air as well as water—trace atmospheric carbon dioxide. We now recognize that both water in the soil and carbon dioxide in the air are principal ingredients for photosynthetic organisms. (Ironically, it was Van

Helmont himself who discovered carbon dioxide gas, but he did not realize its central role in plant growth.)

Even so, the role of sunlight was enigmatic, and it took another three hundred years for the details to emerge. Advances in nuclear physics paved the way, as a new generation of particle accelerators called cyclotrons provided the first steady supply of the highly radioactive isotope carbon-11—a sensitive probe of biological reactions. In the late 1930s, Samuel Ruben and Martin Kamen at the University of California at Berkeley exposed plants to carbon dioxide with a carbon-11 "label." In this way, they could use the telltale radioactivity to follow carbon dioxide as it was taken into plant tissues, though the fleeting twenty-one-minute half-life of carbon-11 made these experiments exceedingly difficult.

Ruben and Kamen's 1940 discovery of a way to manufacture carbon-14, a much more suitable tracer isotope with a leisurely half-life of 5,730 years, revolutionized biophysical research and led to a rapid understanding of how plants take advantage of water, carbon dioxide, and sunlight. In brief, a clever (and very ancient) protein called RuBisCo—a chemical found in the type of pioneering cyanobacteria thought to date back three billion years or more—concentrates carbon dioxide and water and facilitates the assembly of these raw materials into essential bio-building blocks. In the photosynthesis reaction that yields the oxygen we breathe, algae or plants consume six molecules of carbon dioxide plus six molecules of water to make one molecule of the sugar glucose, with six molecules of oxygen as a byproduct. This chemical transformation is another example of our old friend, the redox reaction (like rusting iron). In this case, the carbon atoms in carbon dioxide gain electrons and are thus reduced, while water or some other electron donor is oxidized. In photosynthesis, the Sun's rays provide the energetic boost to shift electrons.

As straightforward as the bare-bones chemical reaction might sound—carbon dioxide plus water (or some other chemical that can contribute electrons) makes sugars and other biomolecules—the details of photosynthesis are immensely complicated and are still being worked out. For one thing, microbes have figured out quite a few different ways to harvest sunlight and other sources of energy. Most oxygen-producing plants and algae today use the bright green pigment chlorophyll to absorb light in red and violet wavelengths. But throughout Earth's history, a variety of cells have employed other photosynthetic pathways that produced no oxygen at all. Alternate light-absorbing pigments have evolved to decorate red and brown algae, purple bacteria, and strikingly beautiful diatoms and lichens in a wide range of colors. A few inventive microbes even power their photosynthetic reactions with infrared radiation—wavelengths that are utterly invisible to our eyes but that our bare skin senses as heat energy.

The complex origins of photosynthesis are the subject of the research of biochemist Robert Blankenship, who holds chaired positions in both the chemistry and biology departments at Washington University in St. Louis. Blankenship and his coworkers, including former colleagues from the influential astrobiology team at Arizona State University, search for signs of early life, both on Earth and on other worlds. Their strategy is to examine the varied photosynthetic pathways of many different kinds of living microbes—purple, brown, yellow, and green—scrutinizing their genomes for similarities and differences. Data come from multiple aspects of the intricate photosynthetic apparatus: the varied nature of photosynthetic pigments, the exact molecular sequences of protein "reaction centers" that shift electrons from one molecule to another, the many ways those transferred electrons are then used to make the cell's building blocks, and even the myriad structures of "antenna systems." (Remarkably, cells have evolved clusters of molecules that operate as tiny light-collecting antennas.)

Blankenship finds that life has devised a bewildering diversity of photosynthetic strategies. Life, it seems, exploits any accessible energy source. Over and over again microbes have figured out new ways to collect light for growth and reproduction—at least five separate pathways, extending deep into Earth's evolutionary history. Many details of that history are obscure, but the most ancient and primitive of these energy-gathering chemical reactions, possibly dating to more than 3.5 billion years ago, clearly produced no oxygen at all. Ancestors of those early cells survive today and illustrate that the most deeply rooted biochemistries were anaerobic, neither requiring nor even tolerating oxygen.

The research of Blankenship and his coworkers not only reveals the wide range of these diverse chemical strategies but also points to a tendency for microbes to shuffle and swap their light-collecting genes, co-opting their rivals' photosynthetic pathways like industrial trade secrets. Indeed, the modern scheme of photosynthesis used by virtually all plants appears to be a combination of two more primitive schemes (prosaically named Photosystem I and Photosystem II). Contemporary organisms can thus piggyback complex biosynthesis reactions and collect and use sunlight far more efficiently than those in earlier stages of life on Earth.

More Oxygen

Even without photosynthesis, Earth's surface would have experienced a leisurely (and correspondingly trivial) oxidation through the slow loss of hydrogen molecules into space. High in the atmosphere, H_2O molecules are vulnerable to the destructive powers of ultraviolet radiation and cosmic rays, which can fragment water into hydrogen plus oxygen. Water's atoms rearrange into other simple molecules, mostly H_2 and O_2, as well as traces of ozone, O_3. The resulting swift-moving hydrogen H_2 molecules, unlike the much heavier lumbering O_2 and O_3 molecules of oxygen, are able to escape the incessant pull of Earth's gravity and fly out into the void of space. Throughout Earth's history, small amounts of hydrogen have been lost in this way, leaving behind a gradual accumulation of excess oxygen. Even today the process continues, as a quantity of hydrogen roughly equal to the atoms in a few Olympic-size swimming pools escapes to space every year. By the same process, smaller Mars, with much less gravity to hold its hydrogen, has shed much of its water. Over 4.5 billion years, most of Mars's near-surface hydrogen has thus escaped to space, while iron minerals near the surface have rusted to give the planet its present red color. Even so, the total amount of oxygen in Mars's thin atmosphere is trivial: were it all to condense onto the surface, the layer of liquid oxygen would be less than a thousandth of an inch thick.

Inexorable oxygen production by hydrogen loss might have turned Earth's surface rusty red over a similar multibillion-year period, but it can't have had much effect on Earth's early environment. Even with the most extreme estimates, less than one atmospheric molecule in a trillion was O_2 before the Great Oxidation Event. (Today it's one in five.) That trivial amount of oxygen would have been snapped up, as fast as it could be generated, at Earth's surface by huge quantities of iron atoms just waiting to be oxidized in the oceans and in the soils. Even if Earth had remained lifeless and eventually sported reddish weathered zones in older, stable parts of the continents, such a superficial coating of rouge would have been purely cosmetic.

Life, too, may have contributed a small inventory of oxygen prior to photosynthesis. In fact, cells have learned at least four different ways to make oxygen from their surroundings. Oxygenic photosynthesis is the big one today, but other biochemical pathways may have played small roles in ancient times.

Life scavenges energy from its environment any way it can. The easiest way to gain energy while releasing oxygen is to start with a molecule that is already oxygen rich and highly reactive. Thus a number of microbes have learned to exploit molecules of peroxide (H_2O_2, produced by reactions high in the atmosphere) to generate O_2 plus energy. Admittedly, such molecular species would have been scarce before the rise of atmospheric oxygen, and such microbial mechanisms can't have played much of a role in modifying Earth's early environment.

A team of microbiologists in Holland recently reported a more relevant oxygen-producing scenario: they discovered remarkable microbes that gain energy by decomposing nitrogen oxides. Early in Earth's history, these so-called NOX chemicals were produced in small amounts through reactions of nitrogen gas with minerals—during lightning storms, for example. Today, owing to widespread use of nitrogen-rich fertilizers, many lakes, rivers, and estuaries are heavily polluted with NOX compounds, which promote large microbial blooms. The newly discovered microbes are able to decompose nitrogen oxides into nitrogen plus oxygen, then use the oxygen to "burn" natural gas, or methane, and thus enjoy a jolt of energy. Such a clever chemical strategy might prove especially useful on an oxygen-starved world like Mars.

Fossil Evidence

Of all the oxygen-producing mechanisms, photosynthesis is the undisputed champion, but how early did photosynthesis and the production of oxygen begin? Paleontologists, who scrutinize the fragmentary tangible remains of ancient living worlds, see the connections between life past and life present more vividly than any other scientists. Perhaps it's not surprising, therefore, that they were among the first to find evidence for an oxygenated Earth more than two billion years ago. In their search for the earliest photosynthesis, the fossil hunters naturally focus on Earth's oldest rocks.

Fossil evidence for ancient photosynthetic cells is spotty at best. Precious few microbial remains make it through billions of years of burial, heating, squeezing, and chemical alteration. What does survive is cooked and crushed, often in ways that necessitate a colorful imagination to achieve any biological interpretation. Colonies of fossil microbes often appear as little more than scatterings of small black smudges, so it's not surprising that every report of microbes more than two billion years old has been met with cautious skepticism, if not outright ridicule.

For much of the past four decades, one of the most ardent defenders of paleontological rigor has been J. William (Bill) Schopf, professor of paleontology at the University of California's Los Angeles campus. Based on his studies of increasingly ancient microbial fossils, Schopf has developed a checklist of traits necessary and sufficient to confirm the claim of life. By first focusing on more recent, well-preserved, and unambiguous specimens, Schopf is the scientist who most convincingly pushed the fossil record further and further back, more than three billion years into the remote Archean Eon.

Schopf's criteria are straightforward and reasonable: fossil microbes must come from properly dated sedimentary layers laid down in environments where microbes could have once lived. Fossils must show uniformity of size and shape—consistent spheres or rods or chains, unlike the kind of shapeless black blobs and streaks that are found in many old rocks. Schopf and his students also employed statistics to remove some of the subjectivity inherent in observations of Earth's oldest sedimentary rocks.

This quantitative catalog of essential traits for any suite of microbial fossils served Schopf well. He was able to publish unassailable descriptions of new fossil finds, while raising doubts about some of the more questionable claims of ancient life by competing researchers. His most notable challenge came in 1996, when NASA scientists announced that microbial remains had been found in a Martian meteorite. At a dramatic NASA-hosted press conference in August of that year, Schopf was the lone dissenting voice. With thinly veiled contempt, he pointed out that the Martian "fossils" were much too small, lacked supporting chemical and mineralogical evidence, and were in the wrong kind of rock to boot. (In spite of Schopf's persuasive arguments, President Clinton lauded the discovery, which may have led to a significant bump in NASA funding for astrobiology—money that wound up supporting many of us in the origins game, including Schopf.)

Ironically, Schopf would soon meet the same kind of withering criticism for an earlier claim he'd made in 1993, when he announced his discovery of Earth's oldest microbial fossils from the Apex chert, a rock formation almost 3.5 billion years old in northwestern Australia. Photographs of suggestive elongated black structures with cell-like segmentation seemed compelling enough. The story, published in a high-profile *Science* paper, supplemented the fossil photos with artistic line drawings (to "aid the eye"), which were placed side by side with photos of cyanobacteria, similar-looking modern photosynthetic microbes. Schopf even claimed that his fossils were probably oxygen producers. Within a few years, his most convincing photos had become among the most reproduced paleontological images of all time, adorning numerous textbooks with captions that repeated the "earliest fossil" claims, often with the suggestion that the microbes were photosynthetic.

It's a rule in science that extraordinary claims require extraordinary evidence. It's also true that extraordinary claims usually receive extraordinary scrutiny. All of Schopf's fossil specimens reside at the British Museum in London, preserved as carefully cataloged, thin transparent sections of rock mounted on glass slides. In 2000 Oxford paleontologist Martin Brasier began a detailed reexamination of the Apex chert material and came to a very different conclusion.

Schopf's "thin sections" of Apex chert actually turned out to be rather thick, at least compared with the size of a microbe. Brasier and his colleagues were eventually able to locate most of the tiny objects that Schopf had photographed and published, but they were surprised to realize that many of the photographs were at best misleading. Each of Schopf's now-classic photos represents a single microscopic focal plane—a thin two-dimensional slice through his smudgy three-dimensional black objects. Brasier and his team employed a newer photographic technique that created a three-dimensional montage of images and thus revealed a much more complex story. Only if they set the microscope focus to the exact depth of Schopf's photographs could they reproduce the now-classic images of Apex "fossils." But raise or lower the focus slightly, and what appeared at first to be a convincing, elongated string of microbial cells morphed into a wavy sheet or irregular blob, sometimes with folds, branchings, or squiggles. According to Brasier's observations, the "chains of microbes" are misleadingly chosen cross sections through complex three-dimensional structures that bear little resemblance to anything biological. The embarrassing challenge by Brasier and colleagues, "Questioning the Evidence for Earth's Oldest Fossils," appeared in the March 7, 2002, issue of the prominent periodical *Nature*.

Schopf fought back with his own article, "Laser-Raman Imagery of Earth's Earliest Fossils," published back-to-back with Brasier's in the same issue. Schopf and his colleagues presented new analyses of the Apex chert's carbon-rich black blobs and showed them to have isotopic compositions and atomic structures consistent with biology. He boldly repeated the "oldest fossil" rhetoric, though he seemed to back off his interpretation that the microbes were photosynthetic. Nevertheless, the seeds of doubt had been sown on Schopf's claims, and the bar had been raised in the hunt for the earliest signs of life.

(In a late-breaking sequel, Martin Brasier and his coworkers from Australia now claim that they have found the "oldest fossils"— microbial remnants from the 3.4-billion-year-old Strelley Pool formation, discovered only twenty miles from Schopf's slightly older, but still disputed, smudges. Few observers expect this development to be the last word in the debate.)

The Smallest Fossils

Imagine what happens when a colony of microbes dies. Almost always the tiny bag of chemicals that once was a living cell fragments and disperses; big biomolecules break down into smaller molecular pieces, mostly water and CO_2. Other microbes may eat the tastiest bits, while indigestible molecules dissolve in the oceans, or evaporate into the air, or become trapped in rock. Usually after a few years, nothing is left, for time is not kind to such fragile molecular remains.

Under extraordinary circumstances—if dead cells are quickly buried, if there's no corrosive oxygen around, if the host rock never gets too hot—a few of the hardiest biomolecules can survive, albeit in a rather altered form. Most likely to persist are molecules with a rugged backbone of up to about twenty carbon atoms, sometimes arranged in a simple long chain (perhaps with a few carbon atoms sticking off to the side here and there), sometimes in a group of interlocking rings (not unlike the Olympic logo). These diagnostic biobits are like ultrasmall skeletons. They represent what's left of much larger collections of functioning molecules that have been degraded and stripped of everything but the most resilient core.

If you can find such a molecular skeleton in an ancient sedimentary rock, and if you can be sure that it's not contamination from nearby younger strata or the ubiquitous leavings of more recently deceased cells (from modern subsurface microbes, for example, or even dead skin from your thumb), then you might be able to claim discovery of a chemical fossil—the actual atoms of a once-living microbe. Hence the fascination with Schopf's black blobs in the Apex chert.

Many modern molecular paleontologists lead a fascinating double life. On the one hand, they may choose to endure the rigors of the field geologist, hiking miles across arduous terrain, schlepping hundreds of pounds of promising rock from remote outcrops in baked deserts, frozen tundra, and high mountains. Every year small teams set out for Western Australia, South Africa, Greenland, and central Canada in search of new specimens. Others labor at drilling rigs in the hopes of securing drill cores of pristine ancient rock, uncontaminated by weather and vegetation. Such expeditions can mean months of hardship, danger, and deprivation.

These adventures contrast with months of tedious analysis conducted in ultraclean labs, where the slightest breath or thumbprint can irrevocably contaminate a precious three-billion-year-old rock sample. It takes time and patience, exquisite care, and an arsenal of sophisticated analytical apparatus to extract individual molecules from a rock. One leading practitioner of this twenty-first-century art is Australian paleontologist Roger Summons, who has set up shop at MIT's department of Earth and planetary sciences. There he heads the Summons Lab, a crack team of a dozen molecular fossil hunters who study Earth's oldest rocks.

A dozen years ago, while working at Australian National University, Summons led a group of scientists who made headlines when they studied promising sediments from the 2.7-billion-year-old Pilbara craton in Western Australia. Summons and his colleagues had access to a unique drill core, a sequence almost half a mile long that included a tantalizing section of black, carbon-rich shale—the kind of sedimentary rock most likely to hold molecular fossils. These Pilbara rocks were of special interest because they appeared to be essentially unaltered by heat and uncontaminated by surface life or groundwater. If ever there was a rock where old biomolecules might survive, that was it.

The Australian researchers focused on hopanoids, the elegant class of hardy biomolecules mentioned in chapter 6. Hopanoids play important roles in stabilizing protective cell membranes and, because of their rarity outside of living cells, are perhaps the most convincing of all the molecular biomarkers. Each hopanoid has a distinctive backbone of five interlocking rings—four little hexagons (each defined by six carbon atoms) and a fifth pentagon (with five carbon atoms) on the end. Each ring shares two carbon atoms with its neighbors, for a total of twenty-one backbone carbon atoms in all.

Meticulous studies at Summons's Australian lab led to two high-profile papers, both published in August 1999. The first one in *Science*, with Summons's Ph.D. student Jochen Brocks as first author, described the discovery of distinctive molecules called steranes in the 2.7-billion-year-old Pilbara rocks—what would have been the oldest known molecular fossils, breaking the previous record by a billion years. Discoveries of steranes can reveal a lot about ancient ecosystems, for different species use a number of different kinds of steranes with extra carbon atoms stuck on at various places around the rings. Brocks and his colleagues suggested that the Pilbara steranes were diagnostic of rather advanced cells called eukaryotes—cells that contain a nucleus that houses DNA. At the time of publication, the oldest known eukaryote fossil cells were only about one billion years old, whereas primitive microbes of the type thought to exist before about two billion years ago don't have a nucleus, so this interpretation was met with surprise, if not downright disbelief. If the discovery is true, then only two conclusions are possible. Either eukaryotes appeared much, much earlier than anyone had thought (and life's evolution was correspondingly accelerated), or else steranes evolved much earlier than eukaryotes. In either case, our understanding of life's history would have to be revised.

The second article, published in *Nature* with Summons as lead author, made the equally startling claim that nearby 2.5-billion-year-old black shales from Mount McRae, a modest 3,300-foot peak in Western Australia, contain a variant of the hopanoid five-ring molecule with an extra carbon atom sticking off the side of the first ring. These 2-methylhopanoid molecules are known only from photosynthetic cyanobacteria, which are Earth's principal oxygen producers. Summons concluded that photosynthesis was well under way on Earth by 2.5 billion years ago. Such a chronology was consistent with the known rise of oxygen at about that time, but the suggestion that the origin of photosynthesis could be sought in preserved molecular fragments opened exciting new doors to paleontology.

Not everyone was persuaded. Like Bill Schopf's earlier claims of "Earth's oldest fossils," Roger Summons's extraordinary hopane findings have been met with some opposition, including grave doubts now raised by Jochen Brocks about his own doctoral work, as well as all other studies of purported biomarkers more than two billion years old. Young hopanoids are everywhere, the skeptics say. The deep subsurface is teeming with microbes living in the rocks, so contamination during more than two billion years of Earth history is unavoidable. The hopanoids and other biomolecules are no doubt there, but who can say when or how they got there. Stay tuned: such debates are fun to watch, and they almost always lead to new discoveries.

Sand Flats of Time

Where else is a paleontologist to look? Of the many clues in the fossil record related to the history of photosynthesis, microbial mats may be at once the most obvious and the most overlooked. Today they form, the world over, in shallow coastal waters and along the banks of slow-moving rivers and streams, where algae can intertwine filaments in thick, tangled layers. These tough, clothlike mats ensure that the algae have access to a wet, sunlit environment, while being protected from the inevitable eroding action of floods and waves. In spite of their widespread distribution, the paleontological community all but overlooked fossil microbial mats before Nora Noffke's discoveries.

For more than a decade, I've had the opportunity to assist Nora Noffke, a professor of geobiology at Old Dominion University in Norfolk, Virginia, and the world's leading authority on ancient microbial mats. Armed with a keen eye, a unique perspective, and steely determination, she has chosen some of the most forbidding areas in the world to conduct her fieldwork. Venturing into remote and hostile places in South Africa, Western Australia, Namibia, the scorching Middle East, and frigid Greenland, she has unearthed paleontological wonders for which no one had previously thought to look. Over and over again Nora has recognized evidence that microbial mats grew on many of Earth's most ancient sandy shores.

The reason microbial mat fossils are so significant is that they must arise from some kind of photosynthesis. The microbes that left their fragmentary remains in black cherts and black shales could have come from deep zones, far from sunlight. A strong case can be made that the shallow-water stromatolites from 3.5 billion years ago supported a photosynthetic lifestyle, though these mineralized mounds could also simply have been protective high-rises in an otherwise harsh, wave-swept environment. But microbial mats must have been photosynthetic. Why would a colony of microbes go to all the trouble of fixing itself in a rough, shallow tidal zone if it wasn't after the sunlight?

To place Nora Noffke's contributions in context, consider other really old fossils. For much of the past half century, paleontologists looking for Earth's oldest life have focused on three kinds of rock formations. First are the black cherts, like Bill Schopf's controversial 3.5-billion-year-old Apex chert. Black cherts first made paleontological headlines in the early 1960s, when Harvard paleobotanist Elso Barghoorn recognized ancient microbial fossils in the 1.9-billion-year-old Gunflint chert from northern Minnesota and western Ontario. Barghoorn scrutinized thin, transparent sections of the fine-grained, silica-rich rock and realized he was seeing ancient microbial body fossils in exquisite detail. With geologist Stanley Tyler, who had first observed enigmatic spherelike objects in the Gunflint a decade earlier, Barghoorn described astonishing suites of unambiguous cells—a microscopic ecosystem of spheres, rods, and filaments, some in the process of dividing. Indeed, in spite of decades of subsequent claims for older fossils, some paleontologists still point to the Gunflint chert as holding the oldest absolutely unambiguous fossils of photosynthetic cells on Earth.

Carbon-rich black shales, a second rock type, of the kind studied by Roger Summons and his colleagues, are perhaps the best source of ancient molecular fossils. Black shales are deep-water accumulations of mud and organic debris, so we can be confident that they entomb the remains of ancient microbes. As a consequence, thick sequences of black shales from Australia, South Africa, and other localities billions of years old are receiving painstaking chemical scrutiny, layer by microscopic layer. As newer, more sensitive analytical tools come online, some capable of detecting single molecules, important discoveries are sure to follow.

The third intensively studied type of ancient fossil-bearing formation is the stromatolites, those layered domelike structures of minerals deposited by early life. Paleontologists might have been stumped by the origin of these mounds, usually preserved in limestone, were it not for modern living stromatolite reefs in shallow seas, most famously in the scenic and remote World Heritage Site of Western Australia's Shark Bay. These odd sedimentary features arise when a slimy surface coating of microbes—photosynthetic microbes, in the case of today's living reefs—produces layer upon layer of minerals. Hundreds of fossil stromatolite localities have been identified around the world, some in rocks older than three billion years.

Black chert, black shale, and stromatolites. To this short list of Earth's oldest fossiliferous formations, Nora Noffke has added a fourth rock type: sandstone. It's understandable why sandstones were overlooked. Most fossils are preserved in fine-grained rocks like chert or shale, or in limestone reefs—hence the focus on black chert, black shale, and stromatolites. Sand, by contrast, is relatively coarse, with mineral grains much larger than most microbes. What's more, sand tends to concentrate at the beach, in the turbulent tidal zone, where most signs of life are quickly erased—eroded, washed away, and dispersed. But Noffke has spent two decades studying modern tidal flats and their rich ecosystems and found that tough, fibrous microbial mats impose distinctive structures on shallow, sandy shorelines. They imprint a crinkly texture to the sand surface, not unlike a wrinkled tablecloth; they bind and trap sediment grains in a thick, resilient mass of algal strands; they alter the pattern of ripple marks in the sand; and they fragment in storms, tearing into distinctive geometric chunks and rolling up like little Persian carpets.

Most sandstone outcroppings appear smooth or gently rippled, devoid of anything obviously biological. But once Noffke learned to spot the distinctive wrinkled and cracked surfaces characteristic of fossilized microbial mats in ancient rocks, she spied subtle features almost everywhere she looked. In 1998 she identified the diagnostic crinkly textures on the surfaces of 480-million-year-old rocks of the Montagne Noire in the French Alps. In 2000, after moving to Harvard University for postdoctoral work, she pushed the record even further back, identifying similar patterns in 550-million-year-old rocks of Namibia. The fact that microbial mats existed a half-billion years ago was not particularly newsworthy; all paleontologists would have agreed that microbial mats must have been around to decorate coastal regions much earlier than that. But no one before Noffke had taken the time to scrutinize modern mat systems, then recognize the similar traces preserved as unambiguous fossils in ancient rocks.

In 2001 Noffke made the first of a series of groundbreaking microbial mat discoveries in formations more than three billion years old from South Africa and Australia—a time long before the presumed Great Oxidation Event. Such features are difficult to spot in the overhead glare of the midday Sun, but late in the afternoon at the end of long and often fruitless days of searching, as the sunlight rakes across barren rock, the telltale wrinkled sandstone surfaces stand in stark relief. "The structures seemed to pop out everywhere," she recalls of one thrilling discovery, consummated at the final hour of the final day of one arduous African field excursion.

Nora first came to me in 2000 at the suggestion of her Harvard mentor, paleontologist Andy Knoll. Andy and I had been friends since graduate student days in the 1970s; for a time, our careers had taken us in different scientific directions, but our mutual interest in astro-biology had renewed our conversations. Knoll realized that Noffke's case for ancient mats was based almost entirely on surface features that, while suggestive, at times required a speculative imagination. The average paleontologist, lacking Noffke's extensive experience with modern mats, could easily overlook or dismiss odd ripple marks or wrinkled rock surfaces. So Knoll encouraged her to bolster her case for mats by adding analytical data on the minerals, biomolecules, and isotopes preserved in her distinctively crenulated layers. Maybe traces of ancient carbon or concentrations of characteristic minerals could provide a smoking gun for some of the oldest, yet ambiguous, matlike features. I had worked with other Knoll students, so I got the call.

The very first specimens that Noffke sent proved to be an object lesson in why such analyses are important. She had found thin wiggly black layers in a sandy sediment three billion years old—what would have been the record oldest microbial mat at the time. She needed confirmation that the black features were carbon-rich with life's req-uisite isotopic signature, with about 3 percent less of the heavier carbon-13 isotope compared to the average crust. She had already written a paper for *Science,* ready to submit, awaiting only that single confirmatory value. The rock samples were rushed from Cambridge, Massachusetts, to the Geophysical Laboratory by FedEx under the highest priority. I was under the gun.

Fortunately my colleague Marilyn Fogel, who is the carbon isotope expert at the Carnegie Institution's Geophysical Lab, was willing to help. Marilyn looked at the sample and told me what to do: crush the rock and grind it to a fine powder, put a few micrograms of powder into each of several tiny cups of pure tinfoil, weigh the samples, and fold each foil cup into a tiny ball the size of a BB. These samples and carbon isotope standards were then fed, one by one, into a furnace that vaporizes any carbon-containing compounds to carbon dioxide gas. The gas flows into a sensitive mass spectrometer that separates and measures carbon-12 and carbon-13. It only took a few hours to obtain the telltale ratio.

Nora was hoping for something in the −25 to −35 range, typical of other microbial mats. But the machine spat out a different story. The isotopic ratio was close to 0, a value that had nothing to do with biology. Rather, it was characteristic of inorganic carbon, the kind that rises from the mantle in fluids and is deposited as thin veinlets of black graphite. Bottom line: the black features in Noffke's samples were carbon-rich, but they were unambiguously *not* biological.

With that object lesson in mind, we proceeded quickly to analyze thin, blackened features in many other promising ancient sediments that Nora had accumulated from her varied field areas—from South Africa, from Australia, from Greenland. Time and again we measured carbon isotopes in the −30 range befitting microbial mats and found other convincing evidence that microbes had flourished along the sandy shorelines of ancient Earth more than three billion years ago. And unlike tiny black smudges or traces of biomolecules, you could see Noffke's evidence in the field, at the scale of an outcrop. Her evidence, you could hold in your hand.

But a core question remains: did the mat microbes produce oxygen, or did they use sunlight for simpler photochemistry? Microbes evolved a variety of sun-harvesting strategies, not all of which produce oxygen. So the details of how those three-billion-year-old mat-forming organisms made their livings will remain a hot topic for some time to come.

The Mineralogical Explosion

The grand arc of the story of Earth's oxygenation is widely accepted. Prior to 2.5 billion years ago, Earth's atmosphere was essentially lacking in O_2. The rise of photosynthetic microbes caused dramatic cumulative changes between about 2.4 and 2.2 billion years ago, when atmospheric oxygen rose to greater than 1 percent of today's concentration. This irreversible change transformed Earth's near-surface environment and paved the way for even more dramatic changes.

As the previous accounts demonstrate, the details of this transition have become the focal points of many scientists' careers. In recent years, my longtime colleague Dimitri Sverjensky and I have jumped into the fray with a striking and somewhat counterintuitive claim: most kinds of minerals on Earth are the consequence of life. For centuries, the tacit assumption has been that the mineral kingdom operates independently of life. Our new "mineral evolution" approach, by contrast, stresses the coevolution of the geosphere and biosphere. We suggest that fully two-thirds of the approximately forty-five hundred known mineral species could not have formed prior to the Great Oxidation Event, and that most of Earth's rich mineral diversity probably could not occur on a nonliving world. In this view, such mineral favorites as semiprecious turquoise, deep blue azurite, and brilliant green malachite are unambiguous signs of life.

The reasons for this mineralogical dependence on the living world are simple. These beautiful minerals, along with thousands of other species, are formed in the shallow crust by the interactions of oxygen-rich waters and preexisting minerals. Subsurface waters dissolve, transport, chemically alter, and otherwise modify the upper few thousand feet of rock. In the process, new chemical reactions occur for the first time, producing new suites of minerals. Sverjensky and I have cataloged long lists of the minerals that are generated this way, deriving from copper, uranium, iron, manganese, nickel, mercury, molybdenum, and many other elements. Prior to the rise of oxygen, such mineral-forming reactions simply could not have occurred.

"What of the red planet Mars?" our colleagues ask. Isn't the rusted surface of our planetary neighbor evidence that Mars has been oxidized and could possess a deep mineral diversity similar to Earth's? No, we argue. The crucial difference is that Mars, and presumably other small planets like it, didn't experience the dynamic circulation of oxygen-rich *subsurface* waters that produces Earth's astounding mineral diversity. There may be stores of groundwater on Mars, as recent data have tantalizingly suggested, but that water is frozen. The only reason Mars is red is that it has lost most of its near-surface hydrogen (and thus most of its water). The little bit of oxygen produced by hydrogen loss makes the surface red, like a thin coating of paint a fraction of an inch thick, but that oxygen can't penetrate very deeply into the Martian crust.

Our new framing of Earth's mineralogical past amplified some prior views. In a 2007 *Science* article, provocatively titled "A Whiff of Oxygen Before the Great Oxidation Event?," geochemist Ariel Anbar and his coworkers meticulously documented trace elements that occur in a sequence of 2.5-billion-year-old black shales from Mount McRae in Western Australia. These finely layered sediments, deposited in the offshore environment of an ancient ocean, appear monotonous to the eye but contain chemical surprises when subjected to close scrutiny. Most notably, a thirty-foot-thick section near the top of the shale is significantly enriched in molybdenum and rhenium—chemical elements that don't generally appear in sedimentary rocks unless they are oxidized. In their more oxidized forms, molybdenum and rhenium dissolve easily from igneous host rock, flowing down rivers and into the ocean, where they can be incorporated into black shale on the ocean floor.

Everyone agrees that these enrichments of molybdenum and rhenium are telling us something about erosion 2.5 billion years ago.

Molybdenite, the commonest mineral of molybdenum (and one that often incorporates rhenium, as well), is exceptionally soft and easily abraded. Perhaps molybdenite-bearing granite was exposed on an ancient mountain slope. Perhaps mechanical weathering produced microscopic bits of molybdenite that were carried to the sea and settled to the black muddy bottom—sediments that would become buried and solidified to form the Mount McRae Shale.

Anbar and his team reached a different conclusion; they proposed that a "whiff of oxygen" from early photosynthetic cells was the agent that did the trick. Perhaps a local concentration of slimy green cells created a microenvironment with enough oxygen to mobilize molybdenum and rhenium. After all, we have unambiguous evidence for the global rise of oxygen 2.4 billion years ago, so why not locally 100 million years earlier?

Sverjensky and I counter that there are plenty of ways besides oxygen to move molybdenum, rhenium, and other elements. Common atmospheric molecules containing sulfur or nitrogen or carbon could have done the electron-accepting trick just as well in the absence of any O_2. Such is the nature of scientific debate as new ideas and arguments are met with alternate claims and counterarguments.

Whatever the exact timing of the rise of oxygen, by Earth's two-and-a-half-billionth birthday, its surface had changed once again. The first dramatic changes occurred on land, as Earth rusted. Oxygen-driven surface weathering began to break down iron-bearing granite and basalt into brick-red soils. As the land aged, it subtly shifted hue from predominantly gray and black to the ruddy color of rust. From space, Earth's continents two billion years ago—though still significantly smaller than today's landmasses—might have appeared something like the modern red planet Mars, but with blue oceans and swirling white clouds that provided dramatically colorful contrasts.

Rust was only the most obvious of many profound mineralogical

changes. Our recent chemical modeling suggests that the Great Oxidation Event paved the way for as many as three thousand minerals,
all of them species previously unknown in our Solar System. Hundreds of new chemical compounds of uranium, nickel, copper, manganese, and mercury arose only after life learned its oxygen-producing
trick. Many of the most beautiful crystal specimens in museums—
blue-green copper minerals, purple cobalt species, yellow-orange uranium ores, and others—speak powerfully of a vibrant living world.
These newly minted minerals are unlikely to form in an anoxic environment, so life appears to be responsible, directly or indirectly, for
most of Earth's forty-five hundred known mineral species. Remarkably, some of these new minerals provided evolving life with new
environmental niches and new sources of chemical energy, so life has
continuously coevolved with the rocks and minerals.

Oxygen, a magically transformative element, plays the starring role
in this drawn-out history. Hungry for electrons, oxygen atoms react
vigorously with all manner of minerals, thus weathering away rocks
and forming nutrient-rich soils in the process. When concentrations
of atmospheric oxygen first rose to significant levels more than two
billion years ago, all photosynthetic life-forms lived in the oceans. The
lands were absolutely barren of life. But oxygen paved the way for life's
eventual expansion across the globe.

Today we experience oxygen in the most intimate exchange. With
every breath we take, a tiny portion of the air becomes a part of us, even
as a tiny part of us becomes the air. As days pass, our bodies melt away
and form again in moment-by-moment chemical reactions with oxygen
and life's other essential elements. Our tissues are replaced over and over
again throughout our lives, Earth's finite store of atoms recycling among
air, sea, land, and all its living forms. Most of the atoms that formed
your infant body at birth are now dispersed, as your present atoms
will be again, if you have the good fortune to live a few more years on
this oxygen-rich planetary home.

Chapter 8

The "Boring" Billion

The Mineral Revolution

Earth's Age: 2.7 to 3.7 billion years

∾ Australian geologist Roger Buick, a dynamic, wiry firebrand of the early-Earth science community, once summed up the period sandwiched between the Paleoproterozoic Era (punctuated by the Great Oxidation Event) and the Neoproterozoic Era (which would see globe-spanning glaciers dominate the surface and life begin to evolve in interesting ways) with these stark words: "The dullest time in Earth's history seems to have been the Mesoproterozoic."

That supposedly uneventful time, the billion years between 1.85 billion and 850 million years ago, is the subject of this chapter. This vast interval, dubbed the intermediate ocean (or more sardonically, the boring billion, by some scientific wits), appears to have been a time of relative biological and geological stasis. No obvious dramatic transformative events took place. At first blush, the rock record reveals no epic, game-changing impacts or sudden climate perturbations. The interface between the ocean's more oxidized near-surface layer and the anoxic ocean depths may have gradually gotten deeper and deeper, but no fundamentally new life-forms seem to have emerged; nor is it generally thought that many new rock types or mineral species arose. At least that's the conventional wisdom.

But *boring* is a risky term. I once made the mistake of calling lipids, the rich and varied class of life's molecules that includes fats and oils and waxes, boring. This remark, made during a public lecture and in ignorance of the nuances of lipid chemistry, was a mistake on two counts. First, lipids are, in fact, amazingly diverse. They play all sorts of interesting roles in regulating life's chemical reactions and crafting its intricate nanoscale structures. Lipids divide the insides from the outsides of most living things. Without them, life as we know it would not be possible. The second reason my remark was a mistake was that I made it unwittingly in the presence of an attentive, humorless chemist who had spent her entire research career studying lipids. She rightly took me to task and sent me lots of highly technical literature to set the record straight. My penance was to read these detailed (and somewhat boring) tomes.

The point is that *boring* may speak more to our profound state of ignorance than to any intrinsic dullness. Earth's boring billion may, in fact, be analogous to human civilization's so-called Dark Ages— that dynamic interval of great innovation and experimentation, inexorable and irreversible change, the gateway to the modern living world, yet once largely ignored by scholars. Our self-inflicted ignorance may be self-reinforcing as well. Ambitious students, who seek to establish their academic reputations in the brief temporal window of graduate school and postdoctoral fellowships, are unlikely to focus on a geological era when nothing much is thought to have happened.

But the geological strata of that enigmatic age must contain surprises for the astute scholar. Hints of dramatic transformations must lie hidden in rocks whose story is largely unread. Some of Earth's most valuable ore reserves—vast deposits of lead and zinc and silver from Zambia and Botswana in Africa, from Nevada and British Columbia in North America, and from the Czech Republic and southern Australia—seem to cluster in rocks of that age. Other localities rich in exotic minerals of beryllium, boron, and uranium also appear to have flourished at about that time. Emerging evidence suggests that Earth's continents may have clumped together into a single gigantic supercontinent during the boring billion, then broken apart, then clumped together again in the planet's most majestic surface cycle. And throughout that billion-year interval, abundant microorganisms—beautifully preserved today as fossils—crowded coastal shallows and offshore environments. Surely we have much to learn about Earth's dark ages.

A History of Change

Dramatic change has been the one constant in our saga of Earth's evolution up to now, a couple hundred million years after its two-and-a-half-billionth birthday. The solar nebula coalesced, and the Sun formed. The dust around it melted into chondrules. Chondrules clumped into planetesimals, and planetesimals into the proto-Earth and other terrestrial bodies thousands of miles in diameter. The impact of Theia, the subsequent formation of the Moon, the incandescent magma ocean hardening to a blackened basaltic crust pockmarked by thousands of explosive volcanoes, the hot sea that soon covered almost all the solid surface so that only the tops of the tallest volcanic cones were dry—all these dramatic events transpired within a half-billion years. Even in the less tumultuous two billion years that followed the accrual of Earth's unique ocean, our planet's surface was constantly in flux, as granite emerged from basalt melts and proto-continents grew on the convection cells that drove plate tectonics.

It was on such a dynamic, variable world that life emerged, evolved, and eventually learned to make oxygen. Constant change was Earth's hallmark. Like a precocious artist, our planet had reinvented itself over and over again, at every stage trying something new.

How, then, could our dynamic planet have found itself mired in an aeon of stasis?

The simple answer is that Earth wasn't static even then. Change was incessant, though perhaps not as dramatic as a Moon-forming impact or the Great Oxidation Event. The boring billion saw the invention of distinctive processes that formed new types of rocks and valuable ore deposits, as well as the first appearance of many new mineral species. And most critically, geological evidence from around the world is revealing that it was a time of coordinated global plate movement that would establish new patterns still in place today.

The Supercontinent Cycle

Earth's familiar geography of oceans and continents is ephemeral, geologically speaking. The Americas, Europe, and Africa framing the mighty Atlantic Ocean; the great eastward sweep of the Asian continent; the expansive Pacific Ocean with its plethora of southern islands and the continent of Australia; and the polar world of Antarctica are but a momentary configuration. The stately process of plate tectonics not only forms the continents but also ceaselessly shuttles them across the globe. Land and water have been subject to extreme makeovers time and time again.

An elite band of geoscientists has learned to tease out our world's ancient, alien cartography and has produced remarkable, if approximate, maps of the once and future Earth. They have many clues to work with. First, we know how the continents are moving today— how fast and in what directions. Year by year the Atlantic opens wider, Africa splits in two, and as we watch amazed, India smashes into China, crumpling the impact zone into the jagged Himalayan

Mountains. It all happens in slow motion, of course, but steadily, an inch or two a year; over the span of a hundred million years, even a snail's pace can produce monumental changes. We play the imaginary videotape of Earth's geography backward and forward, and can guess the features of our planet's capricious face. Even as far back as a half-billion years ago, the rich fossil record of animals and plants can help scientists sketch a picture, especially when the flora and fauna of widely separated continents follow divergent evolutionary pathways. The varied marsupials of Australia, for example, and the large flightless birds of New Zealand tell a compelling story of zoological isolation.

Pushing back more than five hundred million years, the picture begins to fade; we must seek other kinds of clues. Of special import is the fossil magnetism locked into volcanic rocks. We tend to think of our planet's magnetic field in terms of a north-south orientation, familiar from the alignment of a compass needle, but it is more complex than that. Magnetic field lines intersect Earth's surface at an angle, called the dip. At Earth's Equator, the dip is close to zero—nearly horizontal—but at higher latitudes, the dip becomes steeper and steeper until at the poles it is almost vertical. Exacting measurements of the ancient magnetic field frozen into a volcanic rock thus can reveal both the north-south orientation and the latitude of the continent when those rocks solidified. Remarkably, such subtle evidence shows that some rocks now at the Equator were once close to Earth's poles and vice versa. Fossil evidence of former tropical lagoons in Antarctica and frozen tundra in equatorial Africa reinforces such findings. The sedimentary rock record adds vital data. Different kinds of sediments accumulate in different environments: shallow seas, continental shelves, tundra, glacial lakes, tidal lagoons, and swamps each host a distinct rock type.

Bolstered with these clues, experts in paleogeography have managed to craft a coherent and defensible picture of Earth back at least

1.6 billion years, well into the boring billion, with informed speculations extending to even deeper time, to the formation of the first continents. It took a long time for plate tectonics to make the original continents. At the tipping point of subduction, at the very fault line where dense slabs of Earth's earliest basaltic crust plunged down into the mantle depths, unsinkable bits of low-density granite islands piled up one after the next to make larger and larger stable, long-lasting landmasses. These ancient chunks of what are now the continents go by the name *craton*, a term derived from the Greek word for "strength."

Cratons are strong; once formed, they last a long time. Earth today preserves perhaps three dozen more or less intact cratons, some as old as 3.8 billion years and ranging in size from a hundred to more than a thousand miles across. These diverse pieces, each evocatively named—Slave and Superior in North America, Kaapvaal and Zimbabwe in Africa, Pilbara and Yilgarn in Australia—have experienced billions of years of migration across the globe. Jumbled together and ripped apart, along with lots of smaller ancient fragments, they survive as the continents' foundation stones. Three such cratons form most of Greenland, while much of central Canada and the northern parts of Michigan and Minnesota comprise a cluster of a half-dozen others. Large portions of Brazil and Argentina are underlain by several cratons, as are big chunks of northern, western, and southern Australia, Siberia, Scandinavia, a large piece of Antarctica, separate regions of eastern and southern China, most of India, and several swaths of western, southern, and central Africa. All of these cratons began to form more than three billion years ago—a time before modern-style plate tectonics, when only a tiny portion of Earth's surface was dry land. Consequently, all cratons carry a precious, if somewhat warped and scrambled, record of Earth's vibrant adolescence.

Cratons are the keys—the Rosetta stones of early-Earth history. The oceans can't help us decipher early Earth. Thanks to the incessant conveyor belt of plate tectonics, which produces new basaltic crust at the ocean ridges and swallows it up again at convergent boundaries, the oldest ocean crust is not much more than two hundred million years old. Anything older than that must be preserved on the continents or not at all.

The peripatetic cratons have an astonishingly complex history. Propelled by the motions of tectonic plates, they have been shuttled about, colliding with one another to form composite cratons and supercratons, which in turn occasionally clumped into single giant landmasses—continents or supercontinents. Each collision produced a new mountain range along the suture zone; each range provides compelling evidence for the ancient assembly of larger landmasses.

The supercontinents, in turn, rifted apart and fragmented into separate, ocean-bounded island continents. Each time a continent rifted, a widening ocean formed between the diverging pieces and a telltale suite of sediments was laid down: shallow-water sandstone and limestone at first, then deeper water mud and black shale. Such sedimentary sequences point to episodes of continental fragmentation. Over and over, supercontinents have been forged and then ripped asunder. It's an immense jigsaw puzzle with an unknown picture, where pieces constantly change their positions and their shapes.

What does all this have to do with the boring billion? Everything. For a period devoid of flashy signs of activity—a thwack-free, treeless time before elaborate flora and fauna arrived in the geological record—we must turn to paleogeographers to comprehend what it looked like. To decipher the details of the cratons' multibillion-year dance across the globe, these geologists trek to the most remote places on Earth, map the rocks, collect samples, and subject them to a battery of laboratory tests.

At the core of every craton are really old rocks, typically three billion years old or more. These fragmentary parcels of Earth's most ancient crust represent, in total, only a small proportion of the planet's continental mass. They have invariably been baked by heat and pressure, altered by the dissolving powers of subsurface waters, and contorted by crustal stresses. Even so, the nature of the original rocks, whether granitelike intrusions or sedimentary layerings, can often be deduced. Moreover, it's fortunate that cratons aren't static. Throughout their histories, new pulses of magma penetrate the old, forming igneous rock bodies in veins and pods. New sedimentary deposits form inland in lakes and rivers, as well as along shallow, sandy coastlines. Distinctive rock types and characteristic structures also form whenever cratons smack together or rip apart—events that suggest the relative motions of two landmasses. Careful studies of these varied younger formations can discern a suite of rock types that spans the entire history of a craton. Then the fun begins.

The younger rocks provide hints about the chronology of craton movements. Igneous rocks hold tiny magnetic minerals that, when they solidify, lock in the orientation of Earth's magnetic field. Careful paleomagnetic studies can identify not only the orientation of the former north and south poles but also the rocks' approximate latitude when they cooled. Those data, while not exactly GPS coordinates, do record the relative positions of cratons through time. Sedimentary rocks complement these data, for they can host telltale clues about climate and ecology. Sediments deposited in rapidly weathering tropical zones differ markedly from those in temperate lakes or the glacial deposits of higher latitudes. Some sedimentary rocks also incorporate tiny grains of magnetic minerals that hold clues to polar positions.

To garner even a vague sense of Earth's changing face, every one of the three dozen cratons is being scrutinized by armies of geologists. Decades of meticulous fieldwork and lab studies are being undertaken. Data from every part of the planet are being integrated. Then all the cratons will have to be juxtaposed like bumper cars on a globe—and the imaginary movie of their wanderings, starting from the known geography of the modern world, will be played slowly backward. Inevitably, that movie becomes fuzzier and more speculative the further back we go. Nevertheless, the picture that is slowly emerging is extraordinary. According to the latest interpretations, Earth has experienced a repeated cycle of at least five supercontinent assemblies and breakups, extending back perhaps three billion years.

The story of Earth's earliest landmasses is still emerging, and more than a few controversies swirl around the topic. No one has had the nerve to draw more than a sketch map of Earth's surface 3 billion years ago, at least not yet, but one well-vetted idea has named the first continent-size landmass Ur, formed about 3.1 billion years ago from earlier scattered cratonic bits of what are now South Africa, Australia, India, and Madagascar. (An even earlier postulated large landmass, Vaalbara, may have existed about 3.3 billion years ago, but evidence is slim.) According to comparisons of paleomagnetic data from all these Ur-forming regions, what are now separate cratons were sutured together for most of Earth history—their global perambulations appear to have been virtually parallel and thus probably linked. Indeed, magnetic data suggest that the continent of Ur persisted for almost 3 billion years and began to split apart only about 200 million years ago.

The earliest true *super*continent, dubbed Kenorland or Superia (after associated rock localities in North America), is thought to have formed about 2.7 billion years ago from Ur and lots of other smaller pieces. Each time one craton collided with another, a suture zone was formed, while epic compressional forces pushed up a new mountain range. A plethora of such features can be determined from rocks 2.7 to 2.5 billion years old, suggesting a sequential growth of the supercontinent. Paleomagnetic data reveal that Kenorland was at low latitude, probably straddling the Equator, for most of its relatively brief existence.

With those early tracts of land came Earth's first episodes of large-scale erosion and the first great pulses of sediments into the shallow ocean margins. Most early-Earth modelers posit an ancient atmosphere quite different from that of today. Oxygen was totally absent, while carbon dioxide levels may have been hundreds or thousands of times greater than in our time. Rain would have fallen as drops of carbonic acid that ate away at the land and transformed hard rock to soft clays. Rivers carried their muddy cargo into the shallow coastal slopes of the encircling oceans, where thick deltalike wedges of soft sediments accumulated.

By about 2.4 billion years ago, about the same time that oxygen began to accumulate in the atmosphere, Kenorland experienced the flip side of supercontinent formation. Geomagnetic data reveal the divergence of Ur from other cratons as Kenorland began its protracted fragmentation. Those cratonic puzzle pieces scattered from the Equator to the poles. Newly opened shallow seas between the diverging pieces led to thick deposits of shallow marine sediments. The supercontinent cycle had begun.

Hail Columbia

The addition of the supercontinent cycle to the annals of geology has made the boring billion a lot less boring. The next supercontinent episode, one more sharply in focus than Kenorland, thanks to younger and much better preserved suites of rocks, began about two billion years ago, at a time when Earth boasted at least five separate continent-size landmasses. Largest among these was the Laurentian supercraton, a conglomeration of at least a half-dozen cratons thousands of miles across that encompasses much of what is now central and eastern North America. (Specialists in ancient landmasses sometimes refer to this clustering of cratons as the United Plates of America.) The original continent Ur soldiered on as the second-largest landmass, separated from Laurentia by a substantial ocean. The much smaller Baltica and Ukrainian cratons, which form the core of what is now eastern Europe, and cratons representing parts of what are now South America, China, and Africa were also large islands approaching continent size. By the time Earth was 1.9 billion years old, these varied lands had collided at convergent plate boundaries, raising new mountain belts and forming a supercontinent variously named Columbia, Nena, Nuna, or Hudsonland. (The name Columbia, based on persuasive geological evidence from the vicinity of the Columbia River, along the Washington-Oregon border, seems to be used most often.) This vast barren land, roughly estimated to have been eight thousand miles long from north to south and three thousand miles wide, incorporated almost all of Earth's continental crust.

The complexities of retroactively arranging thirty-plus cratonic fragments into one extinct supercontinent are daunting. Not surprisingly, more than one model usually vies for acceptance. In the case of Columbia, two rather different stories emerged almost simultaneously in 2002. On the one hand, geochemist John Rogers of the University of North Carolina and his colleague, Indian geologist Santosh Madhava Warrier (based at Kochi University in Japan), proposed that Laurentia, what is now most of North America, formed the core of Columbia. According to Rogers and Santosh, the continent of Ur was sutured to the west coast of Laurentia; portions of Siberia, Greenland, and Baltica were positioned to the north; and parts of what are now Brazil and West Africa lay to the southeast. In the same year, Guochun Zhao of the University of Hong Kong and several colleagues devised a somewhat different configuration, in which Baltica is sutured to the east coast of Laurentia, while eastern Antarctica and China are attached on the west. Given the great age of Columbia and the preliminary nature of these reconstructions, the agreement between the two science teams is remarkably good. Nevertheless, we can anticipate many debates, as postulated craton locations are shuffled and tweaked for decades to come.

In any case, the assembly of Columbia, commencing 1.9 billion years ago, set the stage for the boring billion. Whatever the configurational details of the Columbian supercontinent actually were, we can be fairly confident that much of its interior was hot, desiccated terrain, with absolutely no vegetation and great expanses of rusty desert. Seen from space, Earth would have appeared as a strangely lopsided world, with its one great enreddened landmass surrounded by an even more expansive (and as yet unnamed) blue superocean. With all the continents concentrated together near the Equator, only modest amounts of ice would have decorated the poles. Ocean levels would have been correspondingly high, perhaps high enough to invade some coastal regions with shallow inland seas.

The equatorial Columbian supercontinent is the supposed starting point for the dullest time in Earth history, but what makes it so dull? What does *stasis* really mean—what parameters were stable? Was it global climate and rainfall? Was it the nature and distribution of life? Was it the composition of the ocean or atmosphere? What measurements have been made to establish this alleged stasis? Conversely, what uncertainties remain unaddressed?

Stasis

Most geology graduate students simply ignore rock formations born between 1.85 billion and 850 million years ago. Four years, while working toward a Ph.D., trying to make a splash and land a tenure-track job, is too short a time to spend on a geological era with such a questionable rap. But Linda Kah was not like most graduate students. Her undergraduate mentor at MIT was John Grotzinger, a leader in the study of Earth's oldest rocks from before 2 billion years ago. Her Ph.D. adviser at Harvard was Andy Knoll—the renowned paleontologist who encouraged Nora Noffke in her microbial mat research. Kah couldn't help but notice that Earth older than 1.8 billion years (described by Grotzinger) was strikingly different from Earth younger than 0.8 billion years (described by Knoll). Something interesting must have happened during the boring billion, and Kah was determined to find out what. And so she has devoted herself to understanding the Mesoproterozoic Era—the immense span of Earth history from 1.6 to 1.0 billion years ago—a time encompassing most of the boring billion.

Even if the Mesoproterozoic was in fact a time of stasis, a billion years of equilibrium would be remarkable. Change is the central theme of Earth's story. The oceans and atmosphere, the surface and deep interior, the geosphere and biosphere—all aspects of our planet have changed incessantly over the aeons. How could Earth have experienced a billion years without any dramatic events, with no significant transitions in the near-surface environments, with no great novelty in the living or nonliving world? Was there really a billion-year period when all the feedbacks of climate and life were in such perfect harmonious balance? How could such a thing have happened?

At a leisurely breakfast meeting near the University of Tennessee campus, Linda Kah patiently explains Earth's dramatic and repeated transformations (and, consequently, why the Mesoproterozoic wasn't boring at all). She comes prepared with a stack of blank sheets of 8½-by-11-inch white paper; while she talks, she draws neat illustrative diagrams in contrasting blue and red ink.

"I had these ideas a decade ago," she says, describing the fruits of her grueling fieldwork in the hostile Mesoproterozoic terrains of the Mauritanian desert in northwestern Africa. She'd love to go back now, but a rise in banditry and kidnapping makes such a field excursion problematic, foolhardy. Instead she will be a member of the next Mars rover science team, a safer choice.

Kah's scientific story is devoted to plate tectonics and the mess they make of Earth's past—the constant shuffling, colliding, splitting, and suturing of landmasses that cause our globe to look radically different every hundred million years. Even during the three-hundred-million-year interval that introduced the boring billion—when the supercontinent Columbia persisted more or less intact—plate tectonics didn't stop. A notable feature of supercontinents is that they continue to grow gradually at the edges, as ocean plates plunge under their margins and new volcanoes rise near the coasts. The modern-day expansion of the Pacific Northwest coast, where such majestic volcanoes as Mount Rainier, Mount Hood, and Mount Olympia are still active, is but one recent example of this age-old phenomenon. So it was with the expanding margins of Columbia.

Even more continental crust was added to Earth's inventory when Columbia experienced episodic rifting and breakup into smaller continents and islands. About 1.6 billion years ago—the beginning of the Mesoproterozoic—the splitting off of the continent Ur to the west from Laurentia and the rest of Columbia to the east led to a major intercratonic sea and the deposition of a massive sedimentary sequence reaching thicknesses of more than ten miles. This heroic deposit, called the Belt-Purcell supergroup, today forms prominent outcrops over much of western Canada and the northwestern United States. Thus, even as supercontinents split apart and eroded, new continental rocks were being produced from the old.

This rifting of Columbia into two diverging landmasses had other consequences. Laurentia, Ur, and the other continents were still all centered more or less about the Equator, which means that there were still no continents at the poles, which means that there was still no thick ice buildup at the poles, which means that ocean levels were still relatively high. Indeed, shallow seas embayed vast swaths of Laurentia's new west coast; likely less than a quarter of Earth's surface was dry. For a time perhaps exceeding two hundred million years, Earth's total land area was dramatically reduced, while thick sedimentary deposits were accumulating in shallow waters around the globe—deposits now preserved as a telltale sedimentary record. No ice also means no glaciers. The interval from 1.6 to 1.4 billion years ago holds none of the characteristic glacial remains—piles of ice-rounded cobbles and boulders, sand and gravel—that are found in most other geological ages. So the dull Mesoproterozoic saw a great deal of change, even if those changes were geological "business as usual."

Supercontinent Reprise: The Assembly of Rodinia

The boring billion saw the formation of not one but two supercontinents. The scattered fragments of Columbia drifted apart for perhaps 200 million years, but a continent can diverge on a globe only for so long before it starts converging once again. About 1.2 billion years ago Ur, Laurentia, and other Mesoproterozoic continents began to reassemble into a new landmass called Rodinia (after the Russian word for "motherland" or "birthplace"). The rock record of far-flung localities in Europe, Asia, and North America preserves an associated worldwide pulse of mountain-building events between 1.2 and 1.0 billion years ago; each new mountain range elevated as converging cratons collided and crumpled.

The exact geography of Rodinia is still a matter of debate, but geological and paleomagnetic data, coupled with the arrangement of cratons on today's globe, place significant constraints. Most models situate the entire supercontinent near the Equator, with Laurentia—what is now most of North America—at the center, and large pieces of all the other continents stuck on to the north, south, east, and west. According to several reconstructions, Baltica and chunks of what are now Brazil and West Africa lay to the southeast, with other pieces of South America to the south and fragments of Africa to the southwest, though details of the relative positions of Australia, Antarctica, Siberia, and China are as yet unsettled.

The distinguishing characteristic of Rodinia is its absence of certain kinds of rocks. Unlike any other interval of the past 3 billion years, few if any sedimentary deposits have been preserved from the period between about 1.1 billion and 850 million years ago. This hiatus means there were probably no shallow seas between continents of the type that hosted the 1.6-billion-year-old Belt-Purcell supergroup. The conclusion: all the continents must have fit neatly together. Nor does it appear that there were any large inland seas, of the kind that once flooded all of central North America and laid the sedimentary foundations for the Great Plains some 100 million years ago. By this model, equatorial Rodinia had a hot, dry, desertlike interior, much like Australia today. For almost 250 million years, the sedimentary rock cycle seems to have all but shut down.

Linda Kah makes her points methodically, but it's clear that she is passionate about her chosen geological interval. In spite of the meager rock record near its end, the great time interval from 1.85 billion to 850 million years ago witnessed many notable changes as a consequence of the cratonic dance. Two supercontinents were assembled during the boring billion, each producing a dozen mountain ranges by cratonic collisions. In between those two gatherings of the lands, as the Columbian supercontinent rifted apart, some of Earth's most impressive sedimentary sequences were laid down. Much of Earth's land was submerged and then became dry again. The rates of sedimentation varied by orders of magnitude. Ice caps disappeared and reappeared. That's a lot of change for a "boring" aeon. But there's another side to the story.

The Intermediate Ocean

Whatever the exact geometry of the globe, everyone agrees that the Rodinia supercontinent must have been surrounded by an even larger superocean, a body that has been named Mirovia (after the Russian word for "global"). Geochemists who study Earth's past have reached the conclusion that if the Mesoproterozoic Era was boring, then Miro-via is the principal reason why.

The Great Oxidation Event, which sets the dynamic period from 2.4 to 1.8 billion years ago apart from all others in Earth history, was pri-marily a time of changes in atmospheric chemistry. Earth's atmosphere transformed from having essentially no oxygen to having a percent or two. That's a monumental change as far as the near-surface environ-ment goes, but to Earth's oceans, such a change was insignificant.

The key lies in relative masses. The oceans contain more than 250 times the mass of the atmosphere. Any small change in atmospheric chemistry, even a 1 percent increase in oxygen, takes a very long time to be reflected in the oceans—perhaps about a billion years.

Geochemists who want to understand the history of the oceans scrutinize a host of chemical elements and their isotopes. Prior to 2.4 billion years ago the oceans were rich in dissolved iron, a state that could be maintained only if the water column was utterly devoid of oxidants (which would have caused iron oxides to precipitate) and also low in sulfur (which would quickly lead to the formation of pyrite and other iron sulfide minerals). With the atmospheric changes of the Great Oxidation Event, some of that iron was removed as iron oxides in shallow water, either directly by oxygen or indirectly by reaction with oxidized weathering products from the land. Oxygen in the at-mosphere also led to the rapid weathering and erosion of sulfur-bearing minerals, which flowed into the oceans and consumed more iron. These chemical changes triggered a massive deposition of banded iron formations, or BIFs—the thick ocean-floor sediments with layer upon colorful layer of iron minerals that now constitute most of the world's iron ore reserves. The BIF-making process was gradual, and the oceans held a lot of iron, so BIF deposition continued for another six hundred million years. By the time of the boring billion, the oceans were still anoxic, but they had lost most of their dissolved iron.

Fast-forward a billion years: photosynthetic algae continued to produce oxygen, which began to take over the oceans; by six hundred million years ago, most of Earth's oceans were oxygen rich, top to bottom. What happened in between, the crux of the boring billion, is known as the intermediate ocean.

In 1998 geologist Donald Canfield of the University of Southern Denmark proposed that sulfur, not oxygen, played the major role in Earth's intermediate ocean. (Many scientists now refer to a sulfur-dominated Mesoproterozoic ocean as the Canfield Ocean.) His provocative thesis, entitled "A New Model for Proterozoic Ocean Chemistry," appeared in the December 3 issue of *Nature* (after almost a year's delay by initially reluctant reviewers) and soon transformed the way many of us think about the ocean through deep time.

The core idea is simple. The Great Oxidation Event produced enough oxygen to influence the distribution of many "redox sensitive" elements, including iron, but not nearly enough to oxygenate the oceans. On the other hand, enhanced weathering and oxidation of the land introduced lots of sulfate into the oceans. Hence the intermediate ocean became enriched in sulfur, while it was poor in oxygen and iron, a steady state that persisted for a billion years.

Hanging On

The fossil record reinforces the view of an ever so slowly changing intermediate ocean. Some rocks deposited between 2 and 1 billion years ago preserve microscopic fossils of unprecedented quality. The 1.9-billion-year-old Gunflint chert of North America, the 1.4- to 1.5-billion-year-old Gaoyuzhuang formation of northern China, and the 1.2-billion-year-old Avzyan formation from the Ural Mountains in Russia contain minute fossil microbes so sharp and clear, some in the intimate act of dividing, that they look just like their modern living counterparts. Yet such striking improvement in the quality of some fossils only reflects their less altered history, not anything intrinsically novel about that time in Earth history.

The protracted anoxic, sulfidic intermediate ocean was both good news and bad news for life. On the plus side, the influx of sulfate provided an excellent energy source for some microbes—they made their living by reducing the sulfate to sulfide. Hints from the fossil record, including distinctive molecular biomarkers, sulfur isotope data, and even some well-preserved microbes in chert, all point to a thriving Mesoproterozoic coastal population of green and purple sulfur bacteria. These sulfur-eating microbes, which persist today in some anoxic environments, produce organic sulfur compounds that smell just awful—like a tragically failed septic system.

Linda Kah jokes, "The Mesoproterozoic was the smelliest time on Earth," in a riff on Roger Buick's "dullest" line.

"When was it smelly?" I ask.

"I think it was smelly the whole time," she replies.

The bad news for life was its reliance on nitrogen. Nitrogen gas (N_2) is abundant, constituting 80 percent of today's atmosphere. The problem is that life's biochemistry can't use nitrogen gas; instead, life requires nitrogen in its reduced form, called ammonia (NH_3). Consequently, life has evolved a clever protein, an enzyme called nitrogenase, that converts nitrogen to ammonia. But there's a catch, as pointed out in a clever 2002 *Science* article by Ariel Anbar and Andy Knoll. The nitrogenase enzyme relies on a cluster of atoms containing sulfur plus a metal, either iron or molybdenum, but neither metal was present in the intermediate ocean. The iron had been removed during the formation of BIFs, so that wasn't an option. Molybdenum, on the other hand, is soluble only in oxygen-rich water, as in today's oceans. During the anoxic time of the intermediate ocean, molybdenum was to be found only near weathering coastlines in relatively shallow water—environments where those sulfur bacteria are suspected to have thrived.

So it was that Canfield's seminal paper was followed by a

stream of publications linking Mesoproterozoic geochemistry and paleontology—two disciplines whose practitioners didn't always speak to one another twenty years ago. The conclusion: the intermediate ocean harbored microbial life, but such life could thrive only near the coasts. Sulfur-reducing bacteria coexisted with oxygen-producing algae. For a billion years, life hung on, but there were few biological innovations.

The Mineral Explosion

Enter mineralogy, another field that has long been taught in a manner strangely divorced from the grand story of Earth—as separate from geochemistry and paleontology as those fields had been from each other. It's an inexplicable bias, for everything we know of Earth's distant past comes from evidence locked in minerals. Yet the majority of mineralogists rarely talk about the ages or evolution of their samples. Rather, for more than two centuries, mineralogical research has focused on their static physical and chemical properties. Investigations of hardness and color, chemical elements and isotopes, crystal structure and external form have dominated the literature of my livelihood.

I, too, was once firmly committed to this two-hundred-year tradition. For the first two decades of my research career, I isolated perfect tiny crystals of common rock-forming minerals, squeezed them to unimaginable pressures between two brilliant-cut diamond anvils, zapped the compressed samples with X-rays, and measured subtle changes in their atomic arrangements. My colleagues and I ignored geological time and geographical place, as we cared little about the age or location of our microscopic samples. We called ourselves mineral physicists and allied ourselves with the nonhistorical sciences of chemistry and physics. Had we bought into the subtle prejudice against mere geological "storytelling"?

This mind-set reflects mineralogy's origins in mining and chemistry, colored perhaps by a subliminal belief that the fields of physics and chemistry are more rigorous than the creative, qualitative yarns of the geologists. (Earth scientists often wonder if that bias might have anything to do with why there are Nobel Prizes in physics and chemistry but not in geology.) Consequently, few mineralogists have thought about the astonishing changes in Earth's near-surface mineralogy through time.

When I joined with seven colleagues to publish our article "Mineral Evolution" in 2008, our objective was in large part to challenge this traditional perspective—to reframe mineralogy as a historical science. Our venture into the mineralogical history of Earth, as well as that of other planets in our Solar System and beyond, posits that Earth's mineralogy has evolved through a sequence of stages, each of which saw changes in the diversity and distribution of minerals. Hence the narrative arc of this book, in which planets progress from mineralogical simplicity to complexity, from only about a dozen minerals in the dust and gas that made our Solar System to more than forty-five hundred known mineral species on Earth today—two-thirds of which could not exist in a nonliving world.

It was a highly technical article, published in the specialty journal *American Mineralogist,* usually read only by hard-core professionals. But the international media quickly picked up the proposition that life and minerals coevolved. *The Economist* and *Der Spiegel, Science* and *Nature,* and a handful of popular science magazines all seized on our educated guesses about a planet's changing mineral diversity. *The New Scientist* even published a clever cartoon showing four "stages" of mineral evolution, from a swimming crystal with fins to an "evolved" crystal with a walking stick. What none of them acknowledged was that these provocative guesses were all speculative. Is Mars really limited to five hundred mineral species? Are nonliving worlds really unlikely to exceed fifteen hundred species? Is it true that it took a living, oxygenated world to triple Earth's mineralogical diversity? We had presented these statements as hypotheses; the quest to test them had yet to begin.

And who could have predicted that the most fruitful place to look would be the rocks of the boring billion?

To put quantitative flesh on the bones of the mineral evolution hypothesis, one must examine individual groups of minerals. Fortunately the world boasts experts in many different mineral groups. That's why I contacted Ed Grew, a research professor in Earth sciences at the University of Maine. Ed is a wiry, intense scientist who has devoted his life to the meticulous study of the minerals that incorporate beryllium and boron—rare elements that occasionally concentrate in big, beautiful crystals. He knows all 108 officially approved beryllium minerals like old friends. Each has a character; each plays a geological role. So I asked him to think about their stories through time. When did they first appear? What processes led to their diversification? Have any beryllium minerals become "extinct"? No one had ever tried to answer such questions. It's hard enough to catalog every mineral of a given element, but to tease out when each species first appeared or disappeared is a monumental task. For beryl, the commonest beryllium mineral (most prized in its deep green variety, emerald), there are thousands of localities. It's a daunting challenge to track down the oldest beryl.

After a year of toil, Ed Grew produced a landmark graph showing the cumulative number of beryllium minerals through time based on thousands of reported occurrences. As expected, it took a very long time for the first beryl to appear—almost 1.5 billion years. The element beryllium is present at only about two parts per million in Earth's crust, so it takes time for hot fluids to select and concentrate that trace beryllium into an enriched fluid that can precipitate beryl crystals. For another billion years, only about twenty different beryllium minerals appeared. By our fledgling theory, a big pulse of new minerals should have occurred during the Great Oxidation Event, between 2.4 and 2.0 billion years ago, but that's not what Ed found. Instead, the biggest increase occurred a bit later, more than doubling the number of known species between about 1.7 and 1.8 billion years ago. That span, right at the beginning of the boring billion, was a time of assembly for the Columbian supercontinent. Perhaps beryllium was concentrated into new minerals during the intense mountain-forming events associated with continental collisions.

Ed Grew followed up with an even more impressive survey of the 263 known boron minerals. Tourmaline, most prized in its gorgeous semiprecious red-green variant, is found in some of Earth's oldest rocks, but that was it for nearly half a billion years. In samples from 2.5 billion years ago, a measly twenty or so different boron species— less than 10 percent of the modern total—are recognized. As with beryllium minerals, Ed observed a doubling of the number of boron species in rocks of the boring billion era, this time in an interval between about 2.1 and 1.7 billion years ago—an interval that brackets the formation of the Columbian supercontinent. Again, this rapid increase in mineral diversity raises a lot of questions—about when the postoxidation diversification actually took off, about the assembly of supercontinents, and about the rise of mineralogical novelty during the boring billion.

For our next foray into mineral evolution, we took on the ninety known minerals of the scarce element mercury—a study that further complicates the picture. Like the much more abundant element iron, mercury can occur in three chemical states—as an electron-rich metal (the familiar silvery liquid of old thermometers), as well as in two different oxidized forms. Accordingly, we anticipated a sharp increase in mercury mineral diversity following the Great Oxidation Event, but the picture that emerged is rather different. As with the history of beryllium and boron minerals, the earliest mercury mineral—its commonest ore, the brilliant red mineral cinnabar—took more than a billion years to appear. Additional species followed in pulses: a dozen new minerals during the assembly of Kenorland; more than a half-billion years of stasis; another half-dozen during the assembly of Columbia. Evidently, as continents collide, the resulting mountain-forming episodes unleash a flood of mineralizing fluids—processes that generate new minerals. However, the discovery that such mineralization was restricted to intervals of supercontinent formation was a big surprise.

And then an even bigger surprise: for an epic interval from 1.8 billion to 600 million years ago—a gap exceeding even the span of the boring billion—nothing. Not even during the time of Rodinia's assembly a billion years ago did a new mercury mineral appear. We now suspect that the sulfide-laden intermediate ocean is to blame. Cinnabar, mercury sulfide, is among the least soluble of all ores. Any mercury atoms that washed into the ancient sulfidic seas would have immediately reacted with sulfur to form submicroscopic particles of cinnabar, to slowly settle to the bottom, to thwart further mercury mineralization. Only in the last 600 million years, as the oceans became oxygen rich, as the lands became covered with life, did the population of mercury minerals explode.

Mysteries

So, was the explosion of new minerals a consequence of the supercontinent cycle, one of the boring billion's signature phenomena? Or was it simply a delayed reaction to the rise of oxygen? And what of the element mercury: is the sulfur-rich ocean the whole story? And what new and unexpected results will studies of the other fifty or so mineral-forming elements reveal? What's clear is that there's a lot more to learn, for we've only just recently had our heads turned by the rich subtleties of that billion-year span.

This poorly documented interval, from 1.85 billion to 850 million years ago, shares the inexorable processes of change that have characterized every stage of our planet's evolution. By 850 million years ago, Earth's near-surface environment had changed irreversibly. The increasingly oxygenated ocean margins teemed with algae and other microorganisms, including the stinky sulfur bacteria, while the lands were poised to burst with new life.

If nothing else, the mysterious, not-so-boring billion teaches us that Earth has the potential to settle into stasis, a benign balance of its many competing forces. Gravity and heat flow, sulfur and oxygen, water and life can find and maintain a stable equilibrium for hundreds of millions of years. But there's always a *but*. Give any one of these forces a nudge, and Earth can become unbalanced, reaching a tipping point with consequences difficult to anticipate—rapid changes that may disrupt the near-surface environment in a matter of years.

And that is what happened next.

EARTH'S AGE *(billions of years)*

0	1	2	3	4	4.567
Hadean Eon	Archean Eon	Proterozoic Eon		Phanerozoic Eon

Chapter 9

White Earth

The Snowball-Hothouse Cycle

Earth's Age: 3.7 to 4.0 billion years

∾ The Proterozoic Eon, spanning almost half of Earth's history from 2.5 billion to 542 million years ago, was a long period of sharp contrasts. Its newsworthy first 500 million or so years witnessed the great flourishing of photosynthetic algae and the consequent rise of atmospheric oxygen, the transformation of once iron-rich oceans by the deposition of massive banded iron formations, and the biological innovation of eukaryote cells with their DNA sequestered in a nucleus—cells that were the precursors of all plants and animals.

The middle billion years of the Proterozoic—the so-called boring billion—were a much more plodding, steadily changing time that was also very smelly.

By contrast, the final three hundred million years were perhaps most dynamic of all, with continental breakup and assembly, radical climate swings, epic shifts in ocean and atmospheric chemistry, and the rise of animal life.

I hope I've established that Earth systems are complexly interconnected. Air, water, and land appear to us as separate spheres, which change over very different scales of time. Weather varies daily; oceans change over millennia; rocks cycle over millions of years; supercontinents take hundreds of millions of years to assemble and break apart. Yet every Earth system affects every other in ways both obvious and hidden from view.

A house serves as a useful, if imperfect, metaphor for our home planet. When you are considering buying a house, you want to know many things—when it was built, for example, as well as the age and geometry of its various additions and renovations. You want details about your house's building materials and their installation, from the foundation to the roof. You need to learn about the plumbing system and its source of water, as well as the air handling system—the furnace and air conditioner and their sources of energy. The smart home buyer also asks about potential risks: from fire and carbon monoxide, termites and carpenter ants, radon and asbestos, leaks and mold. Geologists, likewise, study Earth's origins and major transitions, the nature of rocks and minerals, the movement of water and air, the sources of energy, and the risk from geological hazards.

A house also mimics some of Earth's complex behaviors, as different systems are interconnected in sometimes surprising and unexpected ways through negative and positive feedback loops. On a cold winter day, as the temperature inside the house drops below your comfort level, the thermostat responds by turning on the furnace, and the temperature rises. Once the house warms up, the furnace shuts off. On hot summer days, the air conditioner mirrors that response by tripping on if the inside temperature rises too high. Earth, too, operates through many similar negative feedback loops that help the planet maintain its more or less steady conditions of temperature, humidity, and composition at and near the surface. So, for example, warming oceans result in more clouds, which reflect sunlight back into space and cool the oceans. Likewise, rising concentrations of atmospheric carbon dioxide cause global warming, which accelerates rock weathering—a process that gradually consumes excess carbon dioxide and leads in turn to cooling.

Houses also occasionally display reinforcing or "positive" feedback, sometimes with bad consequences. If your heating system fails on a cold winter day, the pipes can freeze and burst, causing cold water to flood your home, making the house even colder and less livable. Many of the uncertainties regarding Earth's changing climate today focus on potential positive feedback loops and their tipping points. Rising sea levels will lead to coastal flooding, which might result in more evaporation and rainfall, which causes more coastal flooding. A warming ocean might cause the widespread melting of methane-rich ice at and beneath the ocean floor, which could add the greenhouse gas methane to the atmosphere and cause even more warming, which could release even more methane. We have only to look at the runaway greenhouse effect of our sister planet Venus, with its thick carbon dioxide atmosphere and 900-degree Fahrenheit surface temperatures, to see the potentially catastrophic effects of unchecked positive feedback.

The boring billion, to the extent that it really was boring, was a consequence of many efficient negative feedbacks that held change in check. In spite of the globe-spanning migrations of landmasses and the repeated assembly and breakup of supercontinents during that lengthy interval, Earth's climate seems to have been fairly stable. There were no great ice ages. The chemistry of the anoxic, sulfur-rich ocean didn't change much; nor did life evolve in any striking new ways. Some new mineral varieties appeared, but no major tipping points altered the air, land, or sea.

All that was about to change with the breakup of Rodinia.

Breakup

In sharp contrast to Earth's enigmatic quiet interval from 1.85 billion to 850 million years ago, the next few hundred million years experienced some of the most remarkably rapid and extreme near-surface fluctuations in our planet's history. About 850 million years ago, most of Earth's many continental masses were still clumped together near the Equator in the dry and utterly lifeless Rodinia supercontinent. The immense ocean of Mirovia, perhaps punctuated by only a few isolated arcs of volcanic islands, surrounded this denuded, rusty-red megacontinent. The inhospitable atmosphere contained only a small fraction of today's oxygen budget, too little to form much of an ultraviolet protective ozone layer. A time traveler with a reliable supply of oxygen and sunblock might have survived along the coast on a bland diet of algae, but life would have been no picnic on that desolate Neoproterozoic world.

Rodinia's unbalanced juxtaposition of land and sea was not destined to last. For most of Earth's history, climate had been moderated by negative feedbacks. It had varied throughout Earth history, to be sure, but the fluctuations seldom reached life-threatening extremes. Beginning about 850 million years ago, however, several changes disrupted the former equilibrium and pushed Earth to a climatic tipping point. The most important was the gradual breakup of equatorial Rodinia. The first rift 850 million years ago was modest, as the Congo and Kalahari cratons (what are now parts of southern Africa) began to separate to the southwest of the otherwise intact supercontinent. About 800 million years ago, a second small rift zone isolated the West African craton, which moved south from the main landmass. Rodinia's fragmentation was in full swing by 750 million years ago, a time when extensive chains of volcanoes and basaltic lava flows reveal major cracks in the crust. The supercontinent split in half, as a great north-south rift zone separated

Ur to the west and a continental cluster of Laurentia, Baltica, Amazonia, and other smaller cratons to the east.

With rifting came thousands of miles of new coastline and associated pulses of rapid coastal erosion. Dynamic sedimentary basins formed in the intercratonic seas and marked an end to the long hiatus in Earth's rock record—that virtual cessation of sedimentary rock deposition that had begun in the Mesoproterozoic Era and lasted for almost a quarter of a billion years. Microbial life flourished in this shifting, fragmenting world. Eroded lands contributed mineral nutrients to photosynthetic algae, which had long been limited by the ocean's meager store of phosphate, molybdenum, manganese, and other essential elements. Paleontologists imagine a time of shallow, sandy tidal zones with thick mats of slimy green filaments and offshore waters choked with smelly algal rafts.

Tectonic events further conspired to alter Earth's oceans, atmosphere, and climate. Atmospheric oxygen rose in part because of the profligate coastal algal blooms, but also because increased production of algal biomass led to the rapid burial of organic carbon. Throughout Earth's history, carbon-rich biomass has been a major consumer of oxygen. The more biomass that decays, the faster oxygen is consumed. (Forest fires represent an unusually rapid enactment of this ongoing oxygen-depleting phenomenon.) By the same token, the faster carbon-rich biomass is buried, the faster oxygen levels rise. But how can we know if biomass was buried? It turns out that limestone, which precipitates carbon-rich mineral layers on the shallow ocean floor, preserves a subtle telltale record.

Carbon isotopes in limestone point to changes in the production rate of algae. Life's essential chemical reactions—the conversion of water and carbon dioxide to sugar in photosynthesis, for example—always concentrate lighter carbon-12 relative to carbon-13. Consequently, the carbon in biomass (that is, algae, dead or alive) is always "isotopically light" compared with inorganic carbon in limestone.

During normal times, when microbial life flourishes and light carbon is depleted from the oceans, limestone displays a correspondingly heavy isotopic signature. And at times of unusually rapid biomass burial, as even more of the lighter carbon isotope is systematically removed from the oceans, the leftover carbon that makes limestone is even heavier on average. Sure enough, limestone deposited along the shores of Rodinia 790 to 740 million years ago is unusually heavy. During that interval, algae must have spread and been buried at an unprecedented rate.

This profligate blooming of life may have had a significant effect on Earth's climate. Microbial life consumes the greenhouse gas carbon dioxide, which is constantly pumped into the atmosphere by volcanoes. In normal times, CO_2 inputs and outputs are balanced, so atmospheric concentrations remain relatively constant, but during the Neoproterozoic pulses of rapid algal growth, carbon dioxide levels may have dropped, thus reducing greenhouse warming.

Another rather convoluted feedback loop related to CO_2 may have also enhanced Earth's cooling. The rifting of Rodinia resulted in thousands of miles of new seafloor volcanoes, which manufactured hot, lower-density ocean crust. Such hot, buoyant crust tended to support shallower oceans than before, so average sea levels rose. It follows that the period after 750 million years ago was probably one of many inland seas. Inland seas mean more evaporation and increased rainfall, which leads to more rapid weathering of exposed rock. But rock weathering rapidly consumes the greenhouse gas carbon dioxide, and lower CO_2 levels can in turn lead to global cooling.

The distinctive positions of continents and oceans just prior to and during Rodinia's breakup may have played an additional role in altering the global climate. Oceans and land contrast sharply in albedo—the ability to reflect or absorb sunlight. The darker oceans have a correspondingly low albedo; they soak up most of the Sun's energy and grow warmer in the process. Dry barren land, by contrast, is much more reflective. A desiccated, desolate supercontinent like Rodinia would have bounced much of the incident sunlight back into space. Such a juxtaposition of polar oceans and equatorial continents would have exaggerated any global cooling event, since the Equator receives much more solar energy than the poles.

Details of such global-scale processes and complex feedback loops are still being resolved, but it's clear that the Neoproterozoic Earth, after its long period of relative stability, was poised for big changes.

Snowball Earth–Hothouse Earth

Three-quarters of a billion years ago, Earth entered a period of climate instability the likes of which have not been seen before or since. It all began with a brutal ice age.

Glaciers leave an unambiguous suite of sedimentary features. Foremost are thick, irregular layers of diagnostic rocks called tillites, which preserve chaotic jumbles of sand, gravel, angular rock fragments, and fine rock flour. Glaciers also leave behind rounded outcrops of bedrock that have been scratched and polished by the slowly advancing ice sheets. Erratic boulders and moundlike moraines add to the evidence, as do finely layered varved sediments, representing seasonal runoff into glacial lakes.

Field geologists around the world have discovered these glacial features in rocks between 740 and 580 million years old just about everywhere they've looked. Indeed, evidence for an abrupt and drastic climate change about 740 million years ago had been accumulating for decades when geologist Paul Hoffman and three colleagues at Harvard and the University of Maryland published their short, electrifying paper "A Neoproterozoic Snowball Earth" in the August 28, 1998, issue of *Science*. Hoffman and his coworkers made the extraordinary leap that at least twice during that interval, Earth had not just experienced an ice age but had completely frozen over, from poles to

Equator. They based their claim in part on meticulous field observations of a sequence of rocks from the Skeleton Coast of Namibia: thick glacial tillite deposits, side by side with paleomagnetic signals that the glaciers were close to the Equator, at about 12 degrees latitude. And these weren't alpine glaciers from high mountains either; the tillites were clearly deposited in shallow coastal waters, at sea level. The climate must have been correspondingly frigid near the Equator. By contrast, during Earth's most recent ice age, advancing glaciers never got farther south than about 45 degrees latitude, and fossil evidence points to a relatively warm tropical zone even during the maximum extent of ice. The Harvard team had unassailable evidence for Neoproterozoic ice accumulations at sea level, close to the Equator. Hence the snowball Earth.

For many scientists reading Hoffman's article in 1998, carbon isotopes provided the smoking-gun evidence for such a sudden and catastrophic change. During the millions of years prior to the first presumed snowball Earth episode—before about 740 million years ago—the rapid growth of algal biomass had concentrated isotopically light carbon. Contemporaneous limestones deposited in coastal waters around the fragmenting Rodinia supercontinent are correspondingly heavy. On the flip side, if microbial productivity slows or stops, the carbon isotopes in limestone must become on average much lighter. That's exactly what Hoffman and his colleagues found—a huge decrease of more than 1 percent in heavy carbon, just before and just after the appearance of glacial deposits about seven hundred million years ago.

The model that emerged relies on nested positive feedback loops, each of which drove Earth to a colder and colder state. One feedback depended on continental weathering—a process that accelerated in hot and humid tropical zones as it pulled more and more carbon dioxide from the air. Another feedback clicked in as massive blooms of photosynthetic algae scavenged even more CO_2 out of the air.

Meanwhile, as Earth's atmospheric greenhouse weakened and its climate cooled, ice caps began to form and grow at the poles. This fresh white ice and snow reflected more sunlight into space, a positive feedback that cooled Earth more rapidly than before. And even as the ice sheets spread to lower and lower latitudes, the still-warm equatorial continent and the fecund algal ecosystem continued to pull more and more CO_2 from the skies. Earth's climate, temporarily skewed out of balance, reached a tipping point as white ice from both poles extended toward the Equator and eventually may have encircled the globe. In the extreme "hard snowball" version of this scenario, the one promoted by Paul Hoffman and his colleagues, average Earth temperatures plunged to −50 degrees Fahrenheit, as an ice mantle up to a mile thick surrounded the globe.

For many millions of years, Earth was encased in ice (or at least slush). The white "snowball" Earth, unable to soak up solar radiation, seemed forever locked in its icy cocoon, as temperatures never rose above freezing. The global ice age shut down almost every ecosystem. Earth's formerly abundant microbial life was all but wiped out. A few hardy microbes persisted, as they have for billions of years, in the perpetual darkness of ocean-floor hydrothermal vents. Other sparse surviving populations of photosynthetic algae must have found their homes in sunlit zones of cracked thin ice or shallow open water near the warm flanks of volcanoes.

How could Earth have possibly recovered from such a long, cold global winter? The answer lies in our planet's much deeper, inexorable churnings. The white veneer of ice and snow could not stop plate tectonics; nor could it slow the incessant exhalation of volcanic gases from hundreds of black cones that poked through the ice. Carbon dioxide, the dominant volcanic gas, once again began to build up in the atmosphere. With the land jacketed in ice, timely removal of CO_2 by rock weathering and photosynthesis all but ceased. Carbon dioxide concentrations gradually rose to levels not seen in more than a billion years, eventually to perhaps several hundreds of times modern levels, triggering a new positive feedback—a runaway greenhouse effect. Sunlight still scattered off the white landscape, but carbon dioxide in the atmosphere bounced that radiant energy right back to the surface, inevitably warming the planet.

As the atmosphere warmed, small patches of equatorial ice melted for the first time in many millions of years. As dark land became exposed, more sunlight was absorbed and the warming accelerated. The oceans, too, began to clear of their white coverings as positive feedbacks between the Sun and the surface caused Earth to become warmer and warmer.

A Case of Gas

Many scientists now suspect that another positive feedback—a mechanism that is also of significant concern to our modern times—may have exacerbated the rapid global warming. Methane (CH_4), the simplest hydrocarbon fuel and what we burn in our homes as "natural gas," is also a greenhouse gas, but one that is, molecule for molecule, much more effective than carbon dioxide in trapping the Sun's energy. For billions of years, methane has accumulated in ocean-floor sediments, probably by two contrasting mechanisms. The first, which is much better documented and thus a lot less controversial, involves microbes that release methane as part of their normal metabolic cycle. These methanogens thrive in anoxic ocean sediments near many known methane reserves, so large natural gas deposits are thought to have formed by the sustained action of these microorganisms.

Recent experiments point to a possible second, much deeper source of methane—a source that doesn't rely on biology at all. Some scientists suggest that in the deep crust and upper mantle, at depths to more than a hundred miles where extreme temperatures and pressures prevail, water and carbon dioxide can react with common iron-bearing minerals to produce methane. Experiments at high temperatures and pressures attempt to mimic these suspected deep-Earth reactions. In an oft-cited 2004 study at the Geophysical Laboratory, postdoctoral fellow Henry Scott mixed two common crustal ingredients, calcite (calcium carbonate—the common carbon-containing mineral of limestone) and iron oxide, with water. Scott sealed these ingredients into a diamond anvil cell and laser-heated the sample to more than a thousand degrees—the same extreme conditions found in the upper mantle. Remember, the neatest thing about the diamond anvil cell is that diamonds are transparent, so you can see the sample as it reacts and changes. Henry Scott watched as tiny bubbles of methane formed in the sample chamber. The hydrogen of water had reacted with the carbon of calcite to form natural gas. Other experiments in Russia, Japan, and Canada have found similar synthesis of hydrocarbons under a range of supposed deep-Earth conditions.

These experiments are potentially important for understanding Neoproterozoic global warming, for methane may have contributed to a particularly strong positive feedback. Much of the methane stored near the ocean floor is trapped in a fascinating compound called methane clathrate—an icelike crystalline mixture of water and gas that forms outcrops on the continental slopes. (This freezing-cold methane ice actually burns with a bright flame—check out the videos on YouTube.) Vast quantities of methane—several times all other known methane reserves combined, by some estimates—are locked into these methane ices, which form when gas rising from below reacts with cold seawater. Significant additional methane ice is locked in Arctic permafrost—soils in Siberia, northern Canada, and other regions that have been frozen for thousands of years.

An extreme positive climate feedback might occur when ocean waters warm even slightly, which causes the shallowest clathrate deposits to melt and release copious amounts of methane gas into the atmosphere. This methane adds significantly to the greenhouse effect, which causes the oceans to warm even more. Some scientists now point to a possible catastrophic Neoproterozoic release of ocean-floor methane as a way to accelerate global warming, perhaps flipping Earth from cold to hot in a matter of decades.

This Neoproterozoic scenario depends strongly on the sources of methane. If microbes produce most of the ocean's natural gas, then clathrate production probably slowed during snowball episodes, and methane release may not have played such a major role in warming. If, on the other hand, a significant amount of methane rises from the hot pressurized mantle, then stores of methane clathrates would have built up continuously throughout any global cooling spell independent of microbial life, triggering a much greater feedback. So what process produces the methane—deep rocks, shallow microbes, or a combination of the two?

The question of deep versus shallow origins of methane may seem straightforward, but it's a problem colored by a long-standing, sometimes heated international controversy in the oil and gas business. Petroleum is formed primarily of hydrocarbon molecules, of which methane is the simplest and most abundant. It's widely assumed that whatever natural processes form methane also play a role in the formation of oil.

On one side of the debate is the Russian-Ukrainian school, fathered in the mid-nineteenth century by the famed Russian chemist Dimitri Mendeleev, best known for his ubiquitous periodic table of the elements. Mendeleev proposed an abiogenic origin for petroleum long before there were experiments to bolster his claims. "The capital fact to note," he wrote, "is that petroleum was born in the depths of the Earth, and it is only there that we must seek its origin." Mendeleev's ideas enjoyed resurgence in Russia and Ukraine in the second half of the twentieth century, and they inform Russia's thriving oil and natural gas industry. Some Russian geochemists still advocate that virtually all petroleum and natural gas derives from deep abiogenic sources. In their view, some productive oil fields are renewable resources, continuously filling from vast mantle reservoirs below.

Such thinking is scientific heresy to most American petroleum geologists, who cite a litany of evidence for an exclusively biological origin for petroleum: oil is only found in sedimentary horizons where life once thrived; oil is loaded with distinctive molecular biomarkers; the isotopic composition of oil is uniquely lifelike; the trace elements also point to a living source. For many North American petroleum geologists, the case is settled: virtually all petroleum and natural gas is biogenic.

The debate, polarized by decades of Russian-American rivalry, was rekindled in North America by the brilliant, pugnacious, far-reaching Austrian astrophysicist Thomas (Tommy) Gold, who taught at Cornell University before his untimely death in 2004. Gold's chief claim to scientific fame, at least within his chosen specialty of astrophysics, was his realization that metronomic radio pulses from deep space, the so-called pulsars, are in fact rapidly rotating neutron stars. (For a time, some astronomers thought these radio signals must originate from distant alien technologies, hence the astronomical designation for all pulsars is LGM—short for "Little Green Men.")

Though Gold ventured into many other spheres of science—from the physiology of hearing to the consistency of the powdery lunar surface—his most notable contribution outside astrophysics was to champion the abiotic origins of petroleum and natural gas. Petroleum looks biological, he argued, simply because a thriving community of microbes—the "deep hot biosphere"—uses the abiotic hydrocarbons as food. Microbes thus overprint abiotic hydrocarbons with their distinctive biochemical markers—hopanes, lipids, and more. Based on this hypothesis, Gold advocated hydrocarbon exploration in unconventional places like igneous and metamorphic rocks. He even persuaded a Swedish company to drill an exploratory well into such hard rocks—a project that yielded intriguing, if ambiguous results (and lost a lot of unhappy investors a lot of money).

If you listen closely to both sides of this argument, it's clear that the answer to the question of hydrocarbon origins is not yet resolved. Tommy Gold was endlessly inquisitive and eager for answers. Shortly before his unexpected death, he came to my lab to lecture on the deep hot biosphere and to discuss a possible collaboration—experiments that might have helped to resolve the matter. The critical question of methane's origins remains unanswered, but it is not unanswerable. What we need is a new, international effort to understand deep carbon.

The Deep Carbon Observatory

Carbon is arguably Earth's most important element. Carbon is the key to understanding Earth's variable climate and environment. Carbon has long been, and continues to be, the central element in our quest for energy. Carbon is also the crucial element of life and, by extension, the core element in the design of new drugs and myriad other products. We need to understand carbon, not just in its well-studied surface cycles of oceans, atmosphere, rocks, and life, but from crust to core.

So it was that in the summer of 2009, the Alfred P. Sloan Foundation and the Geophysical Laboratory launched the Deep Carbon Observatory (DCO), an ambitious ten-year program to study carbon in our planet, especially its chemical and biological roles in Earth's deep interior. Where is the carbon? How much is down there? How does it move, especially to and from the surface? How extensive is the deep biosphere? This interdisciplinary, international effort has already attracted hundreds of researchers from dozens of countries. We have many objectives, from completing a global census of deep microbial life to monitoring the carbon dioxide emissions from every active volcano on Earth. But discovering the origins of Earth's hydrocarbons, from methane to petroleum, is a centerpiece of the DCO's mission. Geochemist Ed Young and his colleague Edwin Schauble, both at the University of California in Los Angeles, think isotopes will be the key to determining whether a seep of methane on the ocean floor was produced by a rock or a microbe. But their theoretical calculations can't be tested with any ordinary measurement of heavy versus light isotopes. Ed Young wants to measure "isotopologs."

Isotopologs are chemically identical molecules that differ in the arrangement of their isotopes. Methane, with one carbon atom and four hydrogen atoms, comes in a variety of isotopologs. About 99.8 percent of all carbon atoms are the lighter carbon-12 variety, while one in every five hundred atoms is the heavier carbon-13 isotope. By the same token, hydrogen comes in a lighter version (technically "hydrogen-1," but always referred to simply as hydrogen) as well as the heavier hydrogen-2 isotope, which is always called deuterium. On Earth, the typical hydrogen-to-deuterium ratio is about a thousand to one. These ratios mean that about one in every five hundred methane molecules holds a carbon-13 isotope, while about four in every thousand methane molecules contain a deuterium.

Trace amounts of either of these two heavy isotopes are hard enough to measure, but that's not what Ed Young and his colleagues are after. They want to measure the elusive doubly substituted methane isotopologs—the roughly one-in-a-million molecule of methane that holds both a carbon-13 *and* a deuterium (denoted $^{13}CH_3D$) or else two deuteriums ($^{12}CH_2D_2$). According to Edwin Schauble's calculations, the ratio of those two rare isotopologs in any given sample of methane should provide a sensitive indicator of the temperature at which the methane formed. Temperature is the key: if a batch of methane formed at temperatures below 200 degrees, then it must be microbial; if it formed at temperatures above 1,000 degrees, then it is most likely abiotic.

The idea looks great on paper. The trouble is, there's not an instrument in the world that can tease out the ratio of $^{13}CH_3D$ to $^{12}CH_2D_2$. Conventional isotope analysis is based on mass spectrometry, the process of separating molecules according to their masses. These two isotopologs differ by less than a hundredth of 1 percent in mass, posing significant problems in resolving one type from the other. What's more, the isotopologs are present at extremely low concentrations that challenge conventional analysis. Ed Young and his colleagues need a new instrument that enhances both mass resolution *and* molecular sensitivity. That's why one of the first actions of the Deep Carbon Observatory was to help fund a $2 million prototype instrument designed specifically to measure the isotopolog ratios of methane. (In a satisfying display of cooperation, the U.S. National Science Foundation, the U.S. Department of Energy, Shell Oil Corporation, and the Carnegie Institution of Washington are also supporting the effort.) It's a risky endeavor. It will take years to build the instrument, years more before we know if it works. But a definitive answer to the question of the sources of deep methane, and the resulting insights into a methane-driven feedback loop that may drastically alter Earth's climate, is well worth the chance.

Cycles of Change

Back on Neoproterozoic planet Earth, at the tail end of the first snowball episode seven hundred million years ago, the tipping point of climate change had been reached. The inevitable rise in carbon dioxide played a big role; the sudden release of methane from clathrates may have contributed as well. In a geological blink of an eye—perhaps much less than a thousand years—the climate lurched. The snowball Earth transformed to the hothouse Earth as temperatures soared to record levels.

For a long time, perhaps thirty million years, a warm climate prevailed, but the hothouse ensured its own demise. The elevated atmospheric concentrations of carbon dioxide gradually fell from their extremes. Some of the greenhouse gas was removed by reactions with rocks. The denuded land, exposed to rainfall laced with corrosive carbonic acid (a consequence of high atmospheric CO_2), weathered rapidly. The influx of mineral nutrients, coupled with the resurgence of sunshine, led to explosive algae blooms that devoured the greenhouse gas. All of these events are duly preserved in the carbon isotope record.

And so for the next 150 million years, Earth cycled between these extremes. Not once, not twice, but at least three times the ice gathered and retreated, the global climate swinging drunkenly from arctic to tropical and back again. The first episode, called the Sturtian glaciation, reached a maximum about 720 million years ago. The Marinoan glaciation followed at 650 million years, and the less severe Gaskiers glaciation occurred at 580 million years. Thick accumulations of rock in a dozen countries reveal details of this dramatic cycle. As ice retreated, the glaciers left behind huge piles of plucked-up boulders and ground-up rock, lumpy tillites and polished rounded bedrock. Soon thereafter thick crystalline deposits of carbonate minerals covered the tillite layers—another telltale sign of warming oceans. The carbonates formed so rapidly in the CO_2-supersaturated seas that giant crystals several feet long blanketed the shallow ocean floor. These hasty droppings speak of a time when Earth's tortured surface had lost its chemical equilibrium—forever abandoning its boring billion stasis.

For a time following Paul Hoffman's 1998 publication about snowball Earth, geologists embraced the frozen-planet scenario, but the bloom is now falling off the rose. Climate modelers have found it difficult to encase all of Earth in ice, for their calculations suggest that even at times of significant cooling, the Equator should remain temperate. Field geologists now find evidence of moving ice, surface waves, and ocean currents during the maximum freeze—signs of at least some open water. For most geologists, the hard snowball has been replaced by a more benign "slushball" scenario, the new model to beat. Hoffman counters that the slush could represent conditions just before or just after the glacial maximum.

How might we tell the difference? One intriguing line of evidence supporting a hard snowball is a striking, short-lived pulse of banded iron formations laid down about the same time as ice is thought to have covered the globe. It's hard to explain such deposits, for the oceans had been stripped of their iron more than a billion years earlier, before the start of the boring billion. How, then, could the oceans become recharged by iron? One model suggests that the snowball episode sealed the ocean, cutting off all oxygen to the ocean water column. Meanwhile seafloor hydrothermal vents continued to pump fresh iron from the mantle into the deep ocean. Gradually iron concentrations rose, only to be rapidly deposited as new banded iron formations when the glacial episodes ended.

Snowball versus slushball: such controversies are nothing new in science, and this one has remained low-key, friendlier than most. Paul Hoffman has retired, and a new generation has taken up the challenge, for the answers still lie hidden in the rocks.

The Mystery of the Ice

A greater mystery remains. The snowball/slushball Earth episodes were by no means the first periods of glaciation on Earth, nor would they be the last, but the three major Neoproterozoic intervals stand out against the rest. To the best of our knowledge, never before and never since has such an extreme cold snap occurred on Earth. Why should that be? How can one brief period of Earth history have been so different from any other?

Two earlier periods of glaciation, both well preserved in the rock record, were evidently a lot less severe. The earliest known ice advance, a relatively brief event revealed by tillite deposits on ancient South African cratons, occurred about 2.9 billion years ago, in the middle of the Archean Eon. That it should have taken so long for Earth's ice caps to expand from the poles is in itself something of a mystery. Earlier in Earth history the Sun was much fainter—only 70 percent of its present radiance in the first few hundred million years, and not more than about 80 percent during the mid-Archean glaciation. With so much less energy coming from the Sun, other warming mechanisms must have been at play. Many scientists point to much higher levels of greenhouse gases—carbon dioxide, methane, and an orange hydrocarbon haze—as chief among the suspected moderating influences. Higher heat flows from Earth's turbulent deep interior, and greater volcanic outputs also must have played a role in moderating climate.

Ironically, Earth's first glacial episode may have resulted in part from too much greenhouse gas. If the atmosphere's methane content rose, then reactions high in the stratosphere would have produced more and more of the big hydrocarbon molecules that may have given early Earth a hazy orange sky. If that haze became too thick, then some of the Sun's energy would have been blocked and Earth would have cooled.

A second, longer cooling episode, marked by extensive glacial deposits between about 2.4 and 2.2 billion years ago, followed the breakup of the equatorial Kenorland supercontinent. Atmospheric modeling suggests that increased weathering and sediment deposition along newly formed coastlines gobbled up much of the carbon dioxide present at the time. Contemporaneously, the rise of oxygen spelled the demise of atmospheric methane, the other important greenhouse gas. The faint Sun (perhaps 85 percent of modern levels) was insufficient to maintain as effective a greenhouse, so a protracted cold period ensued.

For the next 1.4 billion years—almost a third of Earth history, including the boring billion—no trace of an ice age has been found. Earth's climate seems to have remained in remarkable balance, not too hot and not too cold. To explain such a long time when changes were so restrained, we can invoke a litany of possible negative feedbacks, all of which may have contributed to stasis, but it's hard to pinpoint the cause when there's no obvious effect. What we can say for sure is that Earth reached a tipping point roughly 740 million years ago, and the snowball-hothouse cycle ensued.

The Second Great Oxidation Event

The living world was not insensitive to such extreme global changes; for at least the last 3.5 billion years, changes in the geosphere have profoundly affected the biosphere. As Earth vacillated between its hot and cold extremes, exposed and weathered continental shores contributed pulses of essential nutrients to coastal ecosystems. Manganese, required for photosynthesis, was one such vital mineral. Molybdenum (used for processing nitrogen) and iron (employed in varied metabolic roles) were also supplied in abundance. But of all the chemical elements, phosphorus may have been the most important in the Neoproterozoic seas. Phosphorus is essential for all life. It helps to form the backbone of the genetic molecules DNA and RNA, it stabilizes many cell membranes, and it plays a key role in storing and transferring chemical energy in every cell.

The story of phosphorus fascinates Dominic Papineau, my colleague who did his postdoc work at the Geophysical Lab. Papineau's French Canadian roots are quickly revealed by his gentle accent; his passion for Earth's oldest formations is evident in every corner of his rock-strewn office at Boston College. Polished chunks of stromatolites and banded iron formations testify to his many field areas in remote lands.

Papineau realized that in some ecosystems, the extent of microbial growth is directly tied to the amount of available phosphorus. He envisions a time when there was an unprecedented amount of the nutrient flowing into the shallow coastal seas of the Neoproterozoic. Some of the world's largest phosphorite deposits—sediments laid down as phosphorus-rich cells that died and settled to the bottom— are concentrated in the same time intervals as the snowball-hothouse cycles. He has traveled the globe in search of these ancient phosphorite strata—northern Canada, Finland, Africa, and India—to study their distinctive geological settings and fascinating chemistries.

The phosphorus-driven algal blooms pumped atmospheric oxygen to new levels, perhaps breathable concentrations of 15 percent. But paradoxically, rotting clumps of algae settling to the ocean floor would have reacted rapidly with oxygen in the water column, returning the deep oceans to a deadly anoxic state. Thus the resurgence of life fol- lowing the snowball Earth may well have led to a stratified ocean with an oxygen-rich layer near the surface, anoxic waters below. Dominic Papineau also sees strong parallels to today's coastal areas, where large fluxes of phosphate from fertilizer runoff may stimulate similar algal blooms and deep-water anoxic dead zones.

Which returns us to one of the central tenets of mineral evolution: the coevolution of the geosphere and biosphere. Minerals change life, even as life changes minerals. When I began my graduate studies in Earth science four decades ago, biology seemed all but irrelevant to geology. The grand rock cycle was viewed as separate from the cycles of life. When I asked my thesis adviser whether I should take a biol- ogy course as my final elective, he persuaded me to take quantum mechanics instead. "You'll never use biology," he assured me.

Dubious advice, considering that at every phase of Earth's evolution, from the origins of life onward, life has influenced geology and geology has influenced life. In 2006 geochemist Martin Kennedy of the Uni- versity of California's Riverside campus and four coauthors proposed a particularly novel, if speculative, example of this codependence.

Their article, "The Inception of the Clay Mineral Factory," appeared in the March 10 issue of *Science*. According to their clever thesis, the rise of atmospheric oxygen from a few percent to its present level was accelerated by positive feedbacks between microbes and clay minerals.

Clay consists primarily of ultra-fine-grained microscopic mineral bits that soak up water and form sticky, gooey masses. If you've ever gotten your foot or your car stuck in deep, wet clay, you won't soon forget. A principal mode of clay mineral formation is weathering, especially weathering by chemical alteration under the wet, acidic conditions of the late Neoproterozoic. Kennedy and his coworkers suggest that the rapid postglacial weathering of continents produced significantly more clay minerals than before the three great snowball-hothouse cycles. What's more, there is growing evidence that microbial colonies began to colonize the coastal landscape about this time, and microbes can be especially efficient at turning hard rock into soft clay.

One of the most striking properties of clay minerals is their ability to bind to organic biomolecules. An increased production of clay minerals would have sequestered carbon-rich biomass, and as the clay minerals washed into the oceans, they would have sequestered that carbon in thick piles of fine-grained sediments. According to the Kennedy scenario, burial of carbon led to the rise of oxygen, which further accelerated the chemical production of clay minerals on land, which led to even more carbon burial. Hence, the "clay mineral factory" may have contributed directly to the rise of atmospheric oxygen and the evolution of the modern living world.

The Invention of Animals

Hothouse algal blooms, aided by phosphorus and other nutrients, undoubtedly contributed to the sharp spikes in atmospheric oxygen. The clay mineral factory may have amplified the effect. And so by about 650 million years ago, atmospheric oxygen had risen to near-modern levels. Elevated oxygen, in turn, has been implicated in the rise of complex multicellular life, for only with such high levels of oxygen could organisms adopt the active, energy-demanding lifestyles of jellyfish and worms. Indeed, the earliest known multicellular organisms appear in the fossil record about 630 million years ago, just after the second global snowball glaciation.

To understand the rise of animal life in the Neoproterozoic Era, we must first look further back, more than a billion years back, to just before the boring billion. The sparse fossil evidence points to the rise of a wholly new kind of single-celled life about two billion years ago. Prior to that time, all cells seem to have led physically separate, if codependent, lives. But about two billion years ago, according to a revolutionary idea first expounded by biologist Lynn Margulis at the Amherst campus of the University of Massachusetts, one cell swallowed another whole. Rather than digest the engorged cell, the bigger cell co-opted the smaller one in a symbiotic relationship that forever transformed life on Earth.

Margulis is a creative powerhouse and an intellectual omnivore. Her scientific career has been devoted to understanding how groups of organisms interact and coevolve; she sees symbiotic relationships and the sharing of biological inventions as a pervasive theme in life's history. Her ideas have ruffled more than a few feathers, in part because they deviate from the more orthodox Darwinian view of evolution primarily by mutation and selection. In spite of the controversies, Margulis's theory of endosymbiosis is compelling and almost universally accepted today. Modern plants, animals, and fungi consist of cells with many internal structures—mitochondria that act like tiny power plants, chloroplasts that harness the Sun's energy in photosynthetic organisms, the cell nucleus that holds the genetic molecule DNA. These and other "organelles" in complex cells have their own cell membranes and, in some cases, their own DNA as well. Margulis proposed that each of these organelles evolved from earlier, simpler cells that were engulfed and ultimately co-opted to perform specific biochemical tasks. According to our best guess, that transition began to occur about two billion years ago and set the stage for much more complex, multicellular life.

Margulis continues to see life's evolution as driven by symbiosis and the sharing of traits among disparate organisms—a view that she has taken well beyond endosymbiosis (and that at times places her outside the mainstream). One of her recent crusades, beautifully outlined in a lecture at a meeting of geologists in Denver, Colorado, is in support of a controversial idea by British biologist Donald Williamson. In 2009 Williamson proposed that butterflies represent the merging of the genetic material of two very different animals—the wormlike caterpillar and the winged butterfly. The controversy intensified when Margulis used her privilege as a member of the National Academy of Sciences to shortcut the peer-review process and sponsor Williamson's publication in the Academy's prestigious journal, *Proceedings*. Some members were incensed, calling the hypothesis "nonsense," more suited to the *National Enquirer* than a scientific periodical. Margulis countered that Williamson's article is worthy of serious scrutiny and debate. "We don't ask anyone to accept Williamson's ideas," she said, "only to evaluate them on the basis of science and scholarship, not knee-jerk prejudice."

Whatever the eventual outcome of that debate, Margulis's theory of endosymbiosis is now conventional thought. By the Neoproterozoic Era, complex cells with nuclei and other internal structures were well established and were poised to cross a new symbiotic threshold. More than six hundred million years ago, single-celled organisms learned how to cooperate, to congregate, to specialize, and to grow and move in a collective. They learned to become animals.

The earliest fossil evidence for an animal-dominated ecosystem comes from the so-called Ediacaran Period, which began about

635 million years ago, shortly after the second of the three great snowball Earth events. The first distinctly patterned fossils were recognized from 580-million-year-old rocks from Ediacara in southern Australia (hence the name). These soft-bodied animals, possible relatives of jellyfish and worms, left pleasingly symmetrical impressions, like ornately lined pancakes or fancifully striated leaves up to two feet across. Similar fossils have subsequently been found all over the world in rocks between about 610 and 545 million years old. Most remarkably, the 633-million-year-old phosphate-rich Doushantuo formation of southern China holds clumps of microscopic cells interpreted as animal eggs and embryos. These structures, which grew in shallow seas just after the Marinoan glaciation, appear identical in every respect to modern animal embryos.

So it appears that the severe snowball-hothouse cycle ultimately played a central role in the evolution of the modern world. It would even be fair to say that we multicelled organisms owe our existence to that moment eight hundred million years ago when Earth reached a climatic tipping point after more than a billion years in which steady sunlight and heat-trapping CO_2 had kept it warm. When that carbon dioxide was rapidly consumed by the weathering of new equatorial continents, and reflective ice spread from both poles to the Equator, Earth's temperatures plunged for millions of years—until steady CO_2 buildup, perhaps amplified by the rapid release of methane from the ocean floors, triggered an equally rapid runaway greenhouse effect.

Perhaps more than any other events in Earth's history, these wildly erratic snowball-hothouse cycles reveal a planet knocked out of kilter. The flip-flopping Neoproterozoic climate led directly to an unprecedented rise in atmospheric oxygen, a transition that paved the way for the first animals and plants and the colonization of the continents. With such biological innovation, evolving Earth soon became infested with novelties—swimming, burrowing, crawling, and flying creatures boasting ever more extreme habitats and behaviors. Indeed, with the advent of an oxygen-rich atmosphere 650 million years ago, for the very first time in Earth's long history you the time traveler could have stood on the ancient alien landscape and breathed deeply without dying in agony. For the first time, you might have gathered a meager meal of green slime, while avoiding a fatal dose of ultraviolet radiation.

Today we are once again entering a period of dramatic climate change, and positive feedbacks appear to be taking hold. Reflective glacial ice is melting at an accelerating rate, exposing more and more ocean and land to absorb more of the Sun's energy. Trees are being cut and burned, thus pumping more carbon dioxide into the atmosphere while decreasing the size of the critical CO_2 consuming forest. And perhaps most critically, the accelerating release of methane from permafrost and deep ocean ices may raise global temperatures even more, triggering the release of even more methane, tipping the balance. If Earth's past holds any lessons for our time, the Neoproterozoic's story of sudden climate change should appear at the top of the list. For even as its snowball-hothouse shifts opened up new opportunities for evolving life, with each episode of climate reversal, almost every living thing died.

EARTH'S AGE *(billions of years)*

0	1	2	3	4	4.567
Hadean Eon	Archean Eon	Proterozoic Eon		Phanerozoic Eon	

Chapter 10

Green Earth

The Rise of the Terrestrial Biosphere

Earth's Age: 4.0 to 4.5 billion years (the last 542 million years)

෴ Plate tectonics saved Earth from itself. Slowly, relentlessly, Earth's convecting interior propelled the breakup of the sprawling equatorial Rodinian supercontinent into more manageable chunks. Continental masses shifted poleward, liberating the Equator from ice-accumulating lands, moderating the extreme snowball-hothouse cycle. Abundant new photosynthetic algal life also helped buffer the wild fluctuations of carbon dioxide, while raising oxygen concentrations close to modern levels. Earth has not since endured such excesses of global temperature as those that preceded the Phanerozoic Eon.

At least five kinds of changes have been at work on Earth during these last 542 million years. Continents have continued to shift, first closing one ocean to form yet another great supercontinent, then breaking up to form the still-widening Atlantic Ocean. Climate has fluctuated from hot to cold and back again many times, though not to the snowball-hothouse extremes of the Neoproterozoic. Oxygen has enjoyed a third great enrichment event, only to see atmospheric concentrations drop in half and rebound again. Sea levels have also changed repeatedly, dramatically reshaping Earth's coastlines; the rock record reveals countless rises and falls, often by several hundred feet. But most spectacular of all, life has changed and evolved radically and irreversibly. And throughout all these transformations, life and rocks have coevolved.

Earth has always been a planet of change, but the story of the Phanerozoic Eon is much more sharply in focus and seems correspondingly more elaborate and nuanced in its variations, thanks to a more extensive and less altered rock record. The key to this rich story is a wealth of exquisitely preserved fossils, the consequence of life's newfound ability to make durable hard parts: teeth, shells, bones, and wood. Animals and plants turn out to be particularly sensitive to changes in Earth's near-surface environment, and so their fossilized remains record episode after episode of adaptation. Microbes can weather almost any storm; that resilience, coupled with their unhelpful simple shapes and sparseness in the fossil record, means that no obvious mass extinction can be recognized in rocks of the Precambrian, when microbes ruled. But the Phanerozoic Eon is an altogether different story.

Thus, over the last 542 million years, we see Earth in a new light. Not a planet changing leisurely over tens or hundreds of millions of years, but a rapidly evolving world—every hundred thousand years was visibly different from the last. Part of the reason is that we have a more detailed record, but it's also the nature of life. Animals and plants, especially creatures that colonize the lands, respond to Earth's cycles quickly—they evolve fast or they die. As old species die out, new species take their places.

All the World's a Stage

Shifting continents of the past 550 million years continued to provide a changeable stage for the evolution of Earth and its increasingly diverse biota. The story is well understood in its basic outlines—a rather simple play in three acts.

Act One: The beginning of the Cambrian Period, 542 million years ago, found the Proterozoic supercontinent of Rodinia broken into several large and scattered pieces. The largest expanse, stretching from the south pole to beyond the Equator, was the sprawling continent of Gondwana, named for a geologically revealing region of India. All of today's southern continents plus a big swath of Asia were jumbled together in this one giant landmass measuring more than eight thousand miles from north to south. Other post-Rodinia continents, all of which were located in the southern hemisphere, included the core of Laurentia (what is today North America and Greenland) and several other big islands (including much of Europe). A global ocean, all but devoid of land, dominated the northern hemisphere. Over the next 250 million years, plates moved all the continents northward. Laurentia more than doubled its size, by merging first with what would become Europe, then with a significant part of Siberia.

Act Two: About three hundred million years ago, northward-tracking Gondwana collided with Laurentia to form the most recent supercontinent, Pangaea. One of the most spectacular geological consequences of this merging of Gondwana and Laurentia was the closing of the ancient sea between North America and Africa, a collision that birthed the Appalachian Mountains. Today the Appalachians, stretching from Maine to Georgia, seem a relatively benign, well-rounded sort of mountain range. Such gently rolling topography speaks to the power of erosion, for three hundred million years ago their jagged still-rising peaks soared six or seven miles high, rivaling today's Himalayas as some of the mightiest mountains in the history of the world. Lopsided Pangaea concentrated almost all of Earth's dry land on one side of the planet, three-quarters of it located in the southern hemisphere. For one hundred million years the appropriately named superocean Panthalassa (Greek for "all sea") surrounded Pangaea.

Act Three: The opening of the Atlantic Ocean commenced 175 million years ago, when the great Pangaean landmass began to fragment into seven major pieces. First Laurentia and Gondwana fractured, forming the incipient North Atlantic, while the growing continental split stretched ever farther to the northwest and southeast. Antarctica and Australia split off Gondwana and moved south, forming their own island continents. A rift between South America and the west coast of Africa opened the South Atlantic, while India broke off the east coast of Africa and began its 50-million-year journey northward, ultimately to smash into Asia and crumple up the Himalayan Mountains.

Throughout this protracted history, each of the various continental players scurried here and there, forming partnerships and then breaking up, not unlike a human drama. It helps to see this global play unfold: just Google "Pangaea animations." As you watch, remember that the shifting continents imposed other changes on Earth. Greater stretches of coastline promoted more life in shallow waters. Polar landmasses promoted thick ice sheets, which in turn lowered sea level. Life evolved in harsh competition on larger landmasses, but evolution proceeded independently on isolated continents or in widely separated seas. The locations of mountain ranges and oceans altered climate. Throughout history, as today, each of Earth's great cycles has affected every other one.

The Animal Explosion!

For billions of years, the extent of Earth's microbial life had ebbed and flowed in response to climate, nutrients, sunlight, and more. New evidence from shallow-water sediments suggests that the great algal blooms at the end of the Neoproterozoic were more than just a few temporary blips. For the first time, green photosynthetic algae evolved new strategies to achieve a firm footing on swampy land—the continents were finally looking green at the edges, rather than Martian orange against a blue ocean. As atmospheric oxygen soared in concentration, so too did the stratospheric ozone layer—the radiation barrier that effectively shields Earth's solid surface from the Sun's lethal ultraviolet rays. Such a protective blanket was an essential prelude to the rise of a viable terrestrial biosphere of firmly rooted plants and freely roaming animals.

Strangely, it took animal life another hundred million years to crawl fully onto land. For a very long time, most biological innovation took place in the shallow sunlit seas. For forty million years, multicellular jellyfish and worms appear to have dominated the postglacial oceans. Myriad soft-bodied animals, rarely preserved in the fossil record, fed on seafloor detritus and hid in the recesses of minerals laid down by their microbial forebears. For tens of millions of years, an ecological status quo seems to have prevailed.

That status quo was permanently disrupted roughly 530 million years ago by a striking evolutionary trick: many types of animals learned to build their own protective shells out of hard minerals. No one is quite sure how this evolutionary development occurred, though life had already been depositing mineral layers in reeflike stromatolites for billions of years. Somehow, somewhere, following the Gaskiers glaciation 580 million years ago, an unknown animal evolved the exquisite trick of growing its own protective hard parts out of commonplace minerals—most often calcium carbonate or silica. Such an innovation meant a lot in the struggle for survival, for predators would rather eat a soft-bodied morsel than waste energy breaking a tough mineralized exoskeleton. Soon it was make your own shell or die. The resulting fossil record is amazingly rich, marking a time when sedimentary layers became packed with lifelike remains—a time that has been called the Cambrian "explosion."

Explosion is a misleading moniker. This was no sudden transformation; it took many millions of years for "biomineralization" to catch on. A few sponges with hardened spines, preserved in the fossil-rich Doushantuo formation of southern China's Guizhou province, may have learned the trick as far back as 580 million years. By about 550 million years ago, at the tail end of the Ediacaran Period, a variety of wormlike creatures had learned to craft carbonate minerals into tube-shaped protective homes on the ocean floor.

The first recognizable shelly fauna, though small and fragile, appeared around the world in rocks about 535 million years old. (I recall a special undergraduate field trip to Nahant, on the Massachusetts coast just north of Boston, to collect these rare fossils. The brisk sea air, the waves breaking on the picturesque rocky shore, the gorgeous white puffy clouds, and the blue ocean were all memorable—the scrappy weathered fossils, barely visible to the naked eye, not so much.)

The real "explosion" occurred a few million years later, roughly 530 million years ago, when all manner of shelled animals rather suddenly came on the scene. An evolutionary arms race ensued. Armored predators and armored prey assumed larger and larger dimensions. Teeth and claws arose, as did bony protective plating and sharp defensive spines. Eyes became mandatory in the cutthroat world of the teeming Paleozoic oceans. As countless generations of shelled creatures lived and died, their carbonate bioskeletons contributed to massive, resistant limestone layers, which nobly decorate the last half-billion years of Earth history. Stupendous fossil-packed carbonate cliffs and ridges dot the globe, dominating the landscape in dozens of countries, forming the tallest peaks of the Canadian Rockies and the china-white cliffs of Dover, even capping the summit of Mount Everest.

Of all the Cambrian evolutionary innovations, wide-eyed sea

creatures called trilobites are the most prized, the most photogenic. At this point, a disclaimer is required. I love trilobites. I unearthed my first nearly complete specimen not far from my boyhood home in Cleveland, Ohio, at the age of seven or eight, and I've been collecting them ever since. My cache now holds more than two thousand pieces, all of which are being donated to the Smithsonian. (You can see some of the best specimens in the Sant Ocean Hall at the National Museum of Natural History.) So I'm biased.

Though the initial rise of biomineralization was gradual, at 530 million years ago life with hard parts seems suddenly to have been everywhere. All manner of leggy trilobites and striated clams, nutlike brachiopod shells and delicate fanlike bryozoa, porous sponges and horn-shaped corals are preserved in layer after layer of sedimentary rocks across the globe. In rock sequences from Montana to Morocco, you can lay your hand on the exact layer—the exact sliver of history— when this startling invention of bioarmor really took off.

One of the most impressive spots to study the abrupt shift from soft-bodied to shelled animals is near the historic oasis village of Tiout, nestled in the scenic foothills of the Anti-Atlas Mountains in western Morocco. Many thousands of feet of carbonate sediments, standing almost vertically on end and exposed in the steep-walled valley of the Souss River, provide a continuous record of the end of the Ediacaran and the beginning of the Cambrian. Layer upon layer of the thin, reddish-brown limestone is utterly devoid of familiar fossils. You can walk a mile along the river's gravel-strewn bed, which is dry most of the year, and see little more than the occasional suggestion of a worm burrow.

Then suddenly, in a layer of limestone on a hillside above the village—a horizon that, when viewed from a distance, seems no different from those above or below—the fossils appear. The ancient *Eofallotaspis,* perhaps the earliest of all trilobites, marks the very beginning of the Cambrian explosion. In layers a short distance above (that is, younger than) those historic strata, new species are found: the distinctive, elliptical two-inch-long forms of *Choubertella* and *Dagui-naspis*. The latter species is by far the most common, but one of the most productive and accessible outcrops lies smack in the middle of a saint's gravesite, a holy Muslim shrine. The small, round-domed, white structure is surrounded by a low rock wall that's full of trilo-bites. It wouldn't do for visiting geologists to break out hammer and chisel, to disturb the quiet place. Local children seem exempt, how-ever, and they sell the "Tiout bugs" to tourists, tapping on your car window, holding up their freshly exhumed wares.

"Hey mister, one hundred dirhams!" About twelve dollars.

I don't haggle. I buy them all.

Facies Change

For many years, my fossil collecting was focused on the bugs. It's hard to overstate the thrill of cracking open a rock and finding a complete trilobite inside. Fishermen must get a similar buzz when they hook a big fish, and poker players when they draw a full house; for me, it's finding an exquisite animal that's been hidden for five hundred mil-lion years.

For years, the hunt was enough. Then as a senior undergraduate in the spring of 1970, I took my first real paleontology course from the venerable Robert Shrock. For almost four decades, Bob Shrock taught at MIT, and for almost twenty years following World War II, he chaired the MIT department of geology and geophysics. He was a giant in the field, with numerous classic publications, perhaps most notably *Index Fossils of North America,* a massive photographic com-pendium of characteristic species from every geological time since the Cambrian explosion.

Robert Shrock was a gifted teacher with a gentle smile, a natural who imbued his classes with humor and an unabashed passion for his profession. He taught in an avuncular style by telling vivid stories of past times. He told of the chance horseback discovery a century ago of the Burgess Shale in British Columbia, a 505-million-year-old site whose unparalleled soft-bodied fossils were made famous in Stephen Jay Gould's *Wonderful Life*. He described how cute fossils of little frogs were preserved in fine-grained silt in 300-million-year-old tree stumps at Joggins, on Nova Scotia's western coast (the little frogs jumped into the hollowed-out stumps but couldn't jump out). He painted vivid pictures of 90 million years ago, when a vast inland sea covered what is now the Great Plains of the American Midwest—a sea where monster reptiles and squidlike ammonites vied for supremacy.

By a strange twist, my wife, Margee (then a senior at Wellesley College), and I wound up being Bob Shrock's last two students. In the spring of 1970, student protests against the Vietnam War turned angry; classes were disrupted and property destroyed. Given the pervasive distractions, MIT's administration gave students the option to take courses pass-fail and thus skip final exams. Margee and I were the only two students to opt for a grade in paleontology. Our exhausting final exam, administered piecemeal over the space of a week, was to identify every unknown specimen in a tray of a hundred fossils and then *draw the specimen by hand*. Drawing from nature is admittedly a great way to hone observational skills, but I was no artist. Each pencil sketch was a mini-nightmare; it took forever and consumed more erasers than I can recall.

That was Bob Shrock's last paleontology class. The arrival of the eminent seismologist Frank Press as the new department chairman in 1965 had marked a changing of the guard and a swift shift to a more quantitative and physics-based approach to Earth science. Hand-drawn fossils had no place in that modern world, where plate tectonics shifted curricula as surely as it did continents.

Inspired by that final class, Margee and I spent many weekends camping at nearby fossil-rich localities. Over the next few years, we collected fossil ferns in southern Massachusetts, corals in northeastern Pennsylvania, brachiopods in eastern New York, and trilobites in northwestern Vermont. Shrock's course had taught us to see these fossils in a new context. Each type of rock and each suite of fossils told a story of diverse ancient ecosystems.

We learned that at any given time, several different types of rocks—different facies—are being formed, each at a different place and depth of water. Sandstones form nearest the beach in rough, shallow tidal zones. They feature populations of robust fossil clams and snails with thick shells that could withstand the battering surf. Limestone, by contrast, represents ancient coral reefs and thus hosts a rich array of animal life—stalked crinoids, starfish, snails, brachiopods, and other groups that thrive in a protected sunlit lagoon. The many elegant trilobites from reef ecosystems tend to have large eyes that could scan the full 360 degrees of their environment. Farther offshore, black shales accumulate slowly in deep dark water; their fauna commonly include filter feeders and blind trilobites—animals quite different from those of shallower photic zones.

If each outcrop paints a picture of a time and place, then a sequence of rock layers piled one atop the next tells a rich story of change. Particularly dramatic sequences of layered rock types often occur in association with economically valuable (and correspondingly well-studied) coal deposits. Coal, which formed abundantly in swampy coastal zones three hundred million years ago, commonly occurs sandwiched between layers of sandstone, which are in turn bounded by shale. Such a sequence—shale, sandstone, coal, sandstone, shale repeated over and over—implies significant shifts in sea level as it drops, then rises, then drops again, perhaps in response to the retreat and advance of polar ice and glaciers. An inescapable conclusion is that for hundreds of millions of years, ocean depths have repeatedly varied by hundreds of feet.

To modern humans, with our immense coastal cities and vast sea-side infrastructure, the height of the oceans (at least within the ebb and flow of tides) seems to be a fixed aspect of the globe. It's hard to imagine a change of even 10 feet, much less hundreds of feet. But the recent sedimentary record is unambiguous on this point. Within the last few tens of thousands of years, the oceans have been at levels more than 150 feet higher and more than 300 feet lower than today. Without any possible doubt, such changes will come again to radically alter the shapes of continental coastlines. Such is the story told by the rocks and their fossil ecosystems.

Life on Land

The most dramatic terrestrial transformation in Earth history had to await the rise of land plants—an innovation recorded as distinctive, sturdy microscopic fossil spores in rocks as old as 475 million years. Though no body fossils of the delicate, easily decayed vegetation from that time have yet been found, those first true plants were probably not unlike modern liverworts—rootless, ground-hugging descendants of green algae that could survive only in low, wet places. For a span of more than 40 million years, in terrestrial rock formations around the world, decay-resistant spores are the only physical evidence for land plants. Evolution of these hardy green pioneers seems to have been steady but slow.

About 430 million years ago, a significant change in the global diversity of spore fossils points to a marked shift in the distribution of land plants. Over the next 30 million years, the liverwortlike spores became less abundant, while those similar to modern mosses and simple vascular plants assumed dominance. Rocks of this interval from Scotland, Bolivia, China, and Australia also hold the oldest known unambiguous fossils of the plants themselves—fragmentary remains of club mosses and other suspected relatives of modern vascular plants (those with an internal plumbing system of water-filled tubes). Lacking extensive root systems, these short, stubby plants would have been restricted to low-lying, wet areas.

The fossil record improves with time, as early plants became more widespread and robust. By four hundred million years ago, primitive vascular plants had begun to colonize once-barren lands across the globe. They appeared as spindly, leafless, miniature shrubs with greenish, sun-seeking photosynthetic stems and branches that rose just a few inches above the ground. Their roots efficiently penetrated the rocky ground and provided a sturdy anchor, while capillary action distributed water upward.

In spite of their obvious importance to life's colonization of the land, fossilized plants have long played second fiddle to trilobites and dinosaurs. Animals have more dynamic lifestyles as predators and prey, and they seem more varied in form and behavior—more like us. What's more, fossil plants tend to be fragmentary—typically an isolated leaf or stem, a piece of bark or diagnostically patterned wood. Plants lack what University of Chicago paleobotanist Kevin Boyce calls "the satisfying completeness of clams," but they tell amazing stories nonetheless.

I first met Kevin in 2000, when he was an enthusiastic and creative graduate student at Harvard, working with Andy Knoll and thinking about new ways to tease out those stories from some of Earth's oldest fossil plants. A voracious reader and gifted writer, Kevin has a knack for telling stories—he's the kind of scientist who can make the history of plants captivating. But to tell new stories about Earth's oldest plants, Kevin needed new kinds of data about plant fossils. Andy sent Kevin to the Geophysical Laboratory to learn about microanalytical techniques for elements, isotopes, and molecules—techniques never before applied systematically to fossilized plants.

Our first joint research venture focused on remarkably preserved plant fossils from the four-hundred-million-year-old Rhynie chert of

Aberdeenshire, Scotland. The fortunate Rhynie vegetation was saved from rotting when hot springs surrounded their tissues with mineral-rich waters, hermetically sealing and partially replacing buried plant matter with fine-grained silica. A century ago geologists discovered boulders of the chert in a stone fence in the small village of Rhynie. Only after considerable excavation was a small tract of actual bedrock found and quarried. Rhynie chert specimens remain precious and quite difficult to come by, but Kevin Boyce had access to an old Harvard collection of fist-size specimens and thin, glass-mounted, two-by-three-inch transparent polished rock sections in which plant anatomy down to cellular details can be studied in a microscope. These fossils reveal fragments of a bizarre landscape, at once familiar and profoundly alien, covered in stalklike branching plants, each with green photosynthetic stems but no leaves.

Decades ago the effort to extract information from the Rhynie plant fossils was nothing less than heroic. It required the preparation of hundreds of thin slices, each providing a two-dimensional view of a complex three-dimensional object. Imagine taking your favorite flower, embedding it in hard opaque epoxy, and then trying to reproduce the flower's shape by cutting the epoxy into flat slices and reassembling the whole. That's what the pioneering Rhynie paleobotanists had to do. What they found was a miniature suite of odd, spindly leafless plants—the ancestors of our green world.

Kevin Boyce decided to revisit the Rhynie chert, to tease out new information about Earth's ancient flora. His strategy was to analyze newly cut and polished sections of Rhynie fossils, roughly the size and shape of a quarter. We employed an electron microprobe, a machine that maps out the distribution of chemical elements across a polished rock surface like our glass-mounted sections of chert—a machine familiar to mineralogists but rarely used by paleontologists. We hoped to see if any of the original plant material had been preserved. The trick was to tune the microprobe to carbon, an element more common in life than in hard rocks. We were delighted to find that Rhynie chert fossils are loaded with carbon—and isotopically light carbon, to boot, a compelling sign of its biological origin. The carbon distribution beautifully highlights the distinctive tubular structures of these early vascular plants. Our first paper on cellular-scale mapping of a variety of ancient plant fossils, including odd stems and plant spores from Rhynie, appeared in the *Proceedings* of the National Academy of Sciences in 2001.

Kevin's next step was to see if he could pull any biomolecular information from the fossils. Could we find actual molecular bits from the original plant tissues? Kevin Boyce focused on a mysterious twenty-five-foot-tall treelike organism called *Prototaxites,* which had been the largest known living thing on land four hundred million years ago. The fossils of this organism are enigmatic because they seem to lack the same cellular structures of much smaller coexisting plants. Rather, their "trunks" appear to be composed of intricately interwoven tubelike structures. Working with my Geophysical Lab colleagues Marilyn Fogel and George Cody, Boyce was able to extract and analyze unambiguous molecular fragments from several *Prototaxites* specimens—fragments quite different from those of the adjacent plant fossils. His remarkable conclusion: *Prototaxites* was a giant fungus, perhaps the largest toadstool in Earth's history.

Kevin Boyce's research reinforces the conclusions of the paleobotany community. Earth's landscape four hundred million years ago was at last green, but in an utterly alien way. Scrubby, stalklike plants shared the land with towering treelike fungi and a few small insects and spiderlike animals.

Inventing Leaves

You could have survived quite comfortably on Earth four hundred million years ago. There was plenty of oxygen and water. There was food in the form of plants and bugs. There was shelter under the giant stands of *Prototaxites*. But the landscape would have seemed so strange. There were green stems and green branches but absolutely no leaves.

Indeed, the invention of the first tiny energy-trapping leaves took tens of millions of years more—a transformative development that raised the stakes in the plant kingdom's evolutionary struggle for sunlight. The tallest plant with the biggest leaves enjoyed an advantage and so followed the evolution of fanlike ferns, branching limbs, and sturdy woody trunks. By 360 million years ago, forests had emerged as an utterly new terrestrial ecosystem. For the first time in history, Earth's land was emerald green.

In a theme that has been repeated over and over again, rocks co-evolved with this new verdant life. The rise of fast-spreading land plants, some of which achieved giant treelike stature, had profound mineralogical consequences. Weathering rates of many surface rocks, including basalt, granite, and limestone, increased by an order of magnitude as a consequence of roots and their rapid modes of biochemical breakdown. The resultant soils, rich in clay minerals, organic matter, and a host of microorganisms, became deeper and more widespread and provided an ever-expanding habitat for more and larger plants and fungi.

Root systems, though hidden from view, evolved in remarkable ways as well. Most important were new symbiotic relationships between plant roots and vast networks of fungal filaments called mycorrhiza. This astonishing evolutionary strategy affects the great majority of plants you see today; indeed, many plants tend to grow poorly in soils lacking fungal spores. The mycorrhizal fungi efficiently extract phosphate and other nutrients from the soil and pass them to the plant, which in turn provides the fungi with a steady diet of energy-rich glucose and other carbohydrates. It's hard to imagine that subterranean geometry, but the extensive network of a tree's roots and fungal filaments belowground is quite often far larger than the tree we see aboveground.

Animals, too, experienced profound evolutionary advances as edible plants expanded across the landscape. Invertebrates—insects, spiders, worms, and other small creatures—were the first terrestrial explorers. Vertebrates, which initially appeared in the guise of primitive jawless fish about 500 million years ago, underwent more than 100 million years of evolution in the oceans before their first halting attempts to colonize dry land. Vicious, alien-looking armored fish with bony-plated jaws arose 420 million years ago; much more familiar cartilaginous sharks and bony fish appeared and diversified over the next 20 million years. But dry land was utterly devoid of vertebrates.

The recent discovery of 395-million-year-old fossil fish bones in China provided the earliest signs of the evolutionary transition from fins to four-footed land animals. For at least 20 million years, fish flirted with the shallow, sometimes dry coastal environment. A few fish developed primitive lungs and ventured onto land for longer and longer stays, but many millions of years passed before the first bony animal felt fully at home breathing air. The oldest known fossil bones of a four-legged land animal, a walking fish with finlike feet, come from rocks about 375 million years old.

The gradual transition from fish to amphibian has increasingly come into focus over the past two decades—a blossoming period of spectacular paleontological discoveries from China to Pennsylvania. New fossil finds point to a 30-million-year interval of intermediate forms, progressively more suited to land but still retaining distinctively fishlike anatomical features. The first true amphibians appeared about 340 million years ago in the middle of the so-called Carboniferous Period, a time when swampy forests were thriving in low lying areas around the world. These primitive land animals, characterized by broad, flat skulls and equipped with splayed legs, five-toed feet, ears suited for listening in air, and other terrestrial adaptations, were clearly different from their fish ancestors. By the Carboniferous Period, Earth's solid surface had for the first time evolved to a strikingly modern appearance, with dense green forests of tall fernlike trees, dank swamps, and lush meadows populated with an ever-widening cast of insects, amphibians, and other creatures. And thanks to the profound influence of life, Earth's near-surface rocks and minerals had also achieved something akin to their modern state of diversity and distribution.

Not that Earth had achieved anything close to stasis, mind you. Climates waxed and waned, droughts and floods stressed the land, and the odd asteroid impact and supervolcano eruption caused traumas to life the likes of which we may hope never to see. But Earth and its biota have proven unfailingly resilient to such insults. Life always finds a way to adapt to the reality of now—whenever now is.

The Third Great Oxidation Event

By 300 million years ago, Earth's forests were flourishing. Indeed, so much leafy biomass was being produced and buried that a new rock type, carbon-rich black coal (hence the Carboniferous Period), began to form by the pressure-cooking of thick masses of dead plants. One consequence of this organic carbon sequestration was a new pulse of oxygen into the atmosphere, just as in the previous Neoproterozoic oxidation event. The rise in oxygen was gradual, from roughly 18 percent of the atmosphere 380 million years ago, to 25 percent about 350 million years ago, to a remarkable 30 percent or more 300 million years ago. Indeed, by some estimates the atmosphere's oxygen content briefly soared to more than 35 percent, well above modern levels. And these extreme figures aren't entirely guesses: beautiful specimens of Carboniferous-age amber, fossilized tree sap, appear to preserve ancient atmospheric bubbles that still hold 30 percent or more oxygen.

The rise in oxygen had beneficial consequences for animal life. More oxygen meant more energy and increased rates of animal metabolism. Some creatures took advantage of the extra oomph by growing larger—much larger. Most dramatic were the giant insects, exemplified by monster dragonflies with two-foot wingspans. The increased oxygen also enhanced atmospheric density and made flying and gliding that much easier. Other animals undoubtedly migrated to previously uninhabitable higher elevations with thickening air they could now breathe in.

For a span of tens of millions of years, life on the Pangaean supercontinent flourished. The climate was benign, resources were plentiful, and life evolved with abandon. But then rather suddenly, mysteriously, 251 million years ago, life collapsed, in the most calamitous extinction event in Earth history.

The Great Dying and Other Mass Extinctions

For the past 540 million years, the fossil record has piled up. It speaks of profligate biological invention—hundreds of thousands of known fossil species of corals and crinoids, brachiopods and bryozoa, clams and snails, not to mention the vast number of different microscopic animals. Specialists estimate diversity in excess of twenty thousand known species of trilobites, with dozens more described every year. Given that trilobites inhabited Earth for only about 180 million years (between about 430 and 250 million years ago), that's an average of a new trilobite species every few thousand years. Taking all the rich diversity of fossil life into account, several new species must have appeared on average every century for more than 500 million years.

What's not so immediately obvious from the fossil record are a few stark episodes of mass death, the sudden extinction of millions of species. It's relatively easy to spot something new, and paleontologists are not immune to the temptation of describing "the first" or "the earliest" appearance of a significant taxon or trait. The first plant, the first amphibian, the first cockroach, and the first snake (albeit with tiny vestigial hind legs) all have made the fossil news. One recent paper even trumpeted the discovery of Earth's oldest known fossil penis (from a four-hundred-million-year-old spider)—yet another remarkable find from the Rhynie chert.

Loss is harder to recognize in the fossil record. Extinctions require meticulous teasing out of fossil diversity layer by layer, time interval by time interval, across the globe. Decades of effort have paid off in the documentation of five great mass extinctions—five hellish times over the past 540 million years when Earth has suffered the loss of more than half its species. As more data accumulate, it seems there may have been as many as fifteen other less severe mass extinction episodes as well.

It's not easy to document the sudden loss of species from the fossil record. Given the many advances and retreats of the oceans, the opening and closing of shallow seas, the slowing of sedimentation during cool periods, and the irreversible losses owing to erosion, the rock record is spotty and incomplete, like an encyclopedia with many of its pages randomly ripped out, with a few entire volumes lost. It's also often difficult to obtain exact ages of strata and to match up the timings of formations on opposite sides of the globe. So the disappearance of any group of animals might simply reflect a longish gap in the record. Nevertheless, as fossil databases grow and paleontologists around the world compare notes, the largest extinction events tend to stand out against the more normal background of life and death.

The end of the Paleozoic Era, 251 million years ago, witnessed the greatest mass extinction of all. An estimated 70 percent of land species and a whopping 96 percent of marine species vanished—a disastrous global event called the Great Dying. Never before or since in Earth history have so many creatures (including all the trilobites) disappeared forever.

Scientists aren't yet agreed on what caused the Great Dying. It certainly wasn't a simple, single cause like a giant asteroid impact; nor did it occur all at once. Indeed, multiple reinforcing stress factors might have come into play. For one thing, oxygen levels had begun to drop rapidly from their Carboniferous highs of 35 percent; by 251 million years ago, they were back to roughly 20 percent. That's enough oxygen to support complex animal life, but the drop perhaps added stress to animals that had adapted to more profligate, demanding high-oxygen metabolisms. The end of the Paleozoic also saw an episode of global cooling and a modest ice age, with thick ice covering the south polar portions of Pangaea. A consequent large drop in ocean levels would have provided additional stresses by exposing most of the world's continental shelves. Continental shelves are the ocean's most productive biosphere, so the loss of a large fraction of those shallow coastal zones would have restricted the growth of coral reefs and other diverse shallow water ecosystems, constricting the entire ocean food web.

Large-scale volcanism at the end of the Paleozoic Era, almost exactly coincident with the mass extinction 251 million years ago, represents yet another major disruption of Earth's biosphere—another influence of the geosphere on the biosphere. That protracted megaeruption of as much as a million cubic miles of basalt in Siberia, one of the largest volcanic events in Earth history, must have severely compromised Earth's environment. For hundreds of thousands of years, pulses of volcanic ash and dust would have reduced the Sun's input and exacerbated any ice age. The release of huge quantities of toxic sulfur compounds would have led to acid rain and further environmental deterioration.

As if all these environmental insults weren't enough, some scientists point to the collapse of the ozone layer as yet another possible stress factor in Earth's greatest mass extinction. Mutant fossil spores from end-Paleozoic rocks around the world, from Antarctica to Greenland, provide intriguing evidence, if not a smoking gun. Perhaps volcanic emissions from Siberia triggered chemical reactions high in the atmosphere that depleted the ozone layer, opening the window for mutagenic ultraviolet radiation.

Whatever the causes, the Great Dying left a staggering hole in Earth's biodiversity. It took thirty million years to recover, but recover it did. And, in a theme repeated after every extinction event, loss led to opportunity. A new era, the Mesozoic, saw new fauna and flora evolve to fill the vacant niches.

Dinosaurs!

A successful publisher once advised me that if I wanted to sell lots of science books, I should write about one of two popular topics: black holes or dinosaurs. (The publisher even went so far as to include "black holes" in the title of one of my books that had absolutely nothing to do with black holes.)

So here goes. Dinosaurs came on the scene about 230 million years ago as beneficiaries of the end-Paleozoic mass extinction. These fascinating reptiles started slow and small but diversified and radiated into every ecological niche over a span of more than 160 million years. For a time after the Great Dying, dinosaurs competed side by side with large amphibians, but another significant extinction event 205 million years ago, coincident with another megavolcano episode, wiped out most nondinosaur vertebrates. A dinosaur explosion followed.

Dinosaurs are only the most arresting and charismatic of the Mesozoic Era fauna. By far the commonest fossils from the time are the elegantly coiled marine cephalopods called ammonites. If I hadn't grown up in the vicinity of trilobite-rich Paleozoic rocks, if I had been raised instead in the Mesozoic lands of South Dakota, I probably would have collected ammonites. Their shells are stunningly beautiful, with their spiral symmetry and iridescent surfaces. These segmented cephalopods, distant ancestors of the chambered nautilus, feature exquisite shell ornamentations called sutures that once separated each interior chamber from the next. Unlike trilobites, ammonite shells can't provide a realistic picture of the complete animals. The big protruding head, with its large eyes and ten suckered tentacles, has long since decayed. What remains is just the protective armored home of a much more interesting creature. For 160 million years, ammonites evolved and diversified in the Mesozoic seas.

The Mesozoic Era saw many other important biological developments. The flowering plants first appeared then. So did the first true mammals. And as with every other significant chunk of Earth history, there were many changes in geography and topography to accompany these developments in the living world. Pangaea began to break up, and the Atlantic Ocean was born. Atmospheric oxygen levels continued dropping to a dangerously low 15 percent, only to rebound to roughly the present value of 21 percent. Sea levels fell and rose over and over again, though there's no evidence for any significant glaciation during the Mesozoic—nothing to rival even the ice age that ended the Paleozoic.

Fast-forward to 65 million years ago and one of the worst days in Earth history. An asteroid estimated to have been about six miles in diameter collided with Earth near what is now the Yucatán Peninsula. An epic mega-tsunami swept the globe, followed by massive fires that burned across entire continents. Immense clouds of vaporized rock darkened the skies and all but shut down photosynthesis. This cosmic trauma appears to have descended on a world already at risk. In an echo of the end-Paleozoic extinction event, a great series of volcanic eruptions in India may have already been altering Earth's atmosphere and weakening its ecosystems for hundreds of thousands of years. In another echo, a significant sea level drop appears to have exposed much of the continental shelf at about that time, upsetting the ocean's food web and killing off all but eight known ammonite species out of thousands. Reasons for such a sea level change are not at all obvious for there was no ice age at the time. Some scientists speculate that midocean ridges became less active, causing a cooling, contraction, and consequent sinking of the entire ocean floor.

Whatever the causes, individually or in concert, all the dinosaurs except for one minor lineage—the birds—went extinct. The last of the ammonites also died out. The way for evolving mammals was paved. These small, rodentlike vertebrates had become well established in the company of their larger (and therefore doomed) dinosaur brethren, and their survival of the end-Mesozoic extinction gave them footing in almost every ecological niche. Within ten million years of the Indian megavolcano and coincident asteroid impact, mammals had diversified; within fifteen million, early ancestors of whales, bats, horses, and elephants had evolved.

So it was that mass extinctions repeatedly challenged and winnowed life on Earth. The last 540 million years have seen this ebb and flow again and again. But what of earlier times? Were there no mass extinctions prior to 540 million years ago? Here paleontologists are stumped. Prior to the Cambrian explosion, there are almost no diagnostic fossils to record. The statistics of extinction require significant numbers of distinctive organisms like dinosaurs and trilobites; before 540 million years ago, they simply don't exist. Did microbial life go through similar periods of trauma and species loss? There must have been giant asteroid impacts and episodes of destructive volcanism that sterilized significant fractions of Earth's surface. Certainly life was severely challenged during the snowball Earth episodes, perhaps during earlier glaciations as well. There could have been hundreds of mass extinctions stretching back to the very dawn of life. But we may never know from the spotty, microscopic Precambrian fossil record.

The Human Age

For more than 99.9 percent of Earth's existence, there were no humans. We are but an eyeblink in our planet's history.

The recent rise of *Homo sapiens* may be traced back to the rodentlike survivors of that rogue Manhattan-size asteroid of 65 million years ago. Within a few million years of the dinosaurs' demise, mammals had radiated into vacant ecological niches, to fields and jungles, mountains and deserts, air and oceans. Even so, the last 65 million years have not been easy. Many of these strange and wonderful new mammals died in other mass extinctions 56, 37, and 34 million years ago, from causes as yet uncertain.

Humans ultimately descended from the survivors of the last of those catastrophes. Monkeys, the great apes, and us—we all point to a common primate ancestor about 30 million years ago. The first hominids, the evolutionary family that includes primates who walk erect, arose perhaps 8 million years ago in central Africa.

Meanwhile, a resurgence of glaciation that began around twenty million years ago has increased in intensity and frequency. Perhaps eight separate times in the past three million years, ice has spread from the poles to cover great swaths of the high latitudes, reaching as far south as the upper Midwest. Though not as extreme as the earlier snowball Earth episodes, these repeated ice ages were each accompanied by drastic drops in sea level by hundreds of feet. Ice bridges linked Asia and North America, allowing migrations of all manner of mammals, including mammoths, mastodons, and eventually humans to the New World.

These ice ages may have led to another surprising evolutionary twist. According to one intriguing theory, cold temperatures favor the survival of infants who stay close to their mothers for longer periods, as well as infants with bigger heads (the larger the head, the lower the heat loss). Big heads mean big brains, while more time with mother means more time to learn. Perhaps it is not a coincidence that the first human, *Homo habilis,* or "man the toolmaker," appeared shortly after one of these great glaciations, 2.5 million years ago.

Throughout the intervening millennia, it has been the human lot to endure and adapt to repeated change. Frigid ice advances followed by unusually warm "interglacial" periods; droughts followed by floods; great retreats of the seas followed by equally great advances: such cycles were for the most part mercifully gradual, spanning many generations, and nomadic humans had plenty of time to move and survive. Such adaptations are only among the most recent examples of life responding to the changeable Earth.

Indeed, the last half-billion years of Earth history have seen the most astonishing interplay between life and rocks—a coevolution that continues with a vengeance in the age of technological man. Aeons ago rocks, water, and air made life. Life, in turn, made the atmosphere safe to breathe and made the land green and safe to roam. Life turned the rocks into soils that have, in turn, nurtured more life and become home to an ever-widening array of flora and fauna.

Throughout Earth history, the air, the seas, the land, and life have been shaped by Earth's transformative powers: the power of sunlight and Earth's inner heat, the magic of water, the chemical power of carbon and oxygen, and the ceaseless convection of the deep interior and consequent disruptions of the crust through earthquakes, volcanoes, and the incessantly shifting continental plates. In the midst of these forces, our species has proven to be resilient, clever, and adaptable. We have learned technological tricks to shape our world to our will: we mine and refine its metals, fertilize and cultivate its soils, divert and exploit its rivers, extract and burn its fossil fuels. Our actions are not without consequences. Every day, if we are attuned to the dynamic processes of our planetary home, we can experience every facet of its intertwined creative forces. And we can then understand how devastatingly changeable the world can be, and how utterly indifferent it is to our fleeting aspirations.

EARTH'S AGE *(billions of years)*

0	1	2	3	4	4·567
Hadean Eon	Archean Eon	Proterozoic Eon		Phanerozoic Eon	

↑

Chapter 11

The Future

Scenarios of a Changing Planet

Earth's Age: The next 5 billion years

Is the past prologue to the future? For Earth, the answer is yes, and no.

As in the past, Earth will continue to be a planet of incessant flip-flop patterns of change. The climate will become warmer, then cooler, over and over again. Ice ages will return, as will times of tropical extremes. Plate tectonics will persist, shuffling continents, while opening and closing oceans. Giant asteroid impacts and megavolcanoes will once again disrupt life.

But other changes will be new, and many of them will be as irreversible as the first granite crust. Myriad living species will die out, never to be seen again. Tigers, polar bears, humpback whales, pandas, gorillas—all are doomed. It's very possible that humans will die out, too.

Many details of Earth's history are largely unknown, perhaps unknowable. But our planet's rich history, coupled with natural laws, gives us a sense of what is to come. Let's start with the long view and then focus closer and closer on modern times.

Endgame: Five Billion Years from Now

Earth is almost halfway to its inescapable demise. For 4.5 billion years, the Sun has shone steadily, getting slightly brighter through time as it "burns" through its vast store of hydrogen fuel. For another five billion years (more or less) the Sun will continue to generate nuclear energy by fusing hydrogen into helium. That's what most stars do most of the time.

Eventually the hydrogen will run out. Some smaller stars reaching this stage just peter out, shrinking in size while sending out much less energy than before. Had the Sun been such a "red dwarf" star, Earth's ultimate fate would be to freeze solid. Life, if it survived at all, would be limited to a few hardy microbes deep underground, where liquid water could persist.

The Sun won't die in that pitiful way, though, for it is massive enough to have a nuclear backup plan. Remember that every star must balance two opposing forces. On the one hand, gravity pulls the star's mass inward to make as small a sphere as possible. On the other hand, nuclear reactions, like a continuous sequence of inner hydrogen bomb explosions, push outward and try to make the star bigger. The Sun, in its present stately hydrogen-burning phase, has achieved a stable diameter of about 870,000 miles—the size that has persisted for 4.5 billion years and will continue to persist for about 5 billion more.

The Sun is large enough that when the hydrogen-burning phase is finally over, a new, frantically energetic helium-burning phase will begin. Helium, the by-product of hydrogen fusion reactions, can fuse to other helium atoms to make the element carbon, but this new solar strategy will have catastrophic consequences for the inner planets. Because of the more energetic helium reactions, the Sun will swell larger and larger, like a crazy superheated balloon, into a pulsating red giant star. It will swell past the orbit of little Mercury, engulfing the tiny planet. It will swell past the orbit of our neighbor Venus, swallowing that sister world as well. It will swell to more than one hundred times its present diameter—even past the orbit of Earth.

The exact details of Earth's endgame are murky. According to some bleak scenarios, the red giant Sun will simply overwhelm Earth, which will vaporize in the solar atmosphere and be no more. Other models have the Sun shedding more than a third of its present mass in unimaginable solar winds (which would ceaselessly blast Earth's dead surface). As the Sun becomes less massive, Earth's orbit will move outward—perhaps just enough to avoid being engulfed. But if we are not engorged by the expanding Sun, all that would remain of our once beautiful blue world would be an utterly barren orbiting cinder. Sparse subsurface microbial ecosystems may persevere for another billion years, but never again will the land be lush and verdant.

Desert World: Two Billion Years from Now

Ever so slowly, even in its present calm hydrogen-burning state, the Sun is getting hotter. In the beginning, 4.5 billion years ago, the Sun shone with 70 percent of its present light. The Great Oxidation Event 2.4 billion years ago found a Sun shining with perhaps 85 percent of today's intensity. And a billion years from now, the Sun will be brighter still.

For a time, perhaps for many hundreds of millions of years, Earth's feedbacks may moderate the change. More heat means more evaporation, which produces more clouds, which reflects more sunlight back into space. More heat means faster weathering of rocks, which consumes more carbon dioxide, which lowers the amount of greenhouse gases. And so negative feedback loops may keep Earth habitable for a long time.

But there will come a tipping point. Smaller Mars reached that critical time billions of years ago, as almost all its surface water was lost. A billion years from now, Earth's oceans will have begun to evaporate at an alarming rate, and the atmosphere will have become a perpetual sauna. No ice caps or glaciers will remain, as even the poles will become tropical zones. For a span of many millions of years, life may thrive in such a hothouse environment. But as the Sun continues to warm and more water vapor enters the atmosphere, hydrogen will be lost to space at ever-increasing rates, slowly drying out the planet. By the time all the oceans are dry, perhaps two billion years hence, Earth's surface will be barren and baked; life will be on the precipice.

Novopangaea or Amasia: 250 Million Years from Now

Earth's demise is inevitable, but it's a very, very, very long way away. Projections into the less remote future paint a more benign picture of a dynamic yet relatively safe planet. Looking forward a few hundred million years, the past is indeed the key to understanding the future.

Plate tectonics will continue to play a central role in changing Earth. Today the continents are scattered. Wide oceans now separate the Americas, Eurasia and Africa, Australia, and Antarctica from one another. But these landmasses are constantly in motion, at rates of roughly an inch or two per year—a thousand miles every 60 million years. We can establish rather precise vectors for every landmass by studying the ages of ocean-floor basalts. Basalt near the midocean ridges is quite young, just a few million years at most. By contrast, basalt at continental margins and subduction zones may be more than 200 million years old. It's a fairly simple matter to take all these ocean-floor ages, play the plate tectonics tape backward in time, and obtain a glimpse of Earth's shifting continental geography during the past 200 million years. From that information, it may also be possible to project plausible plate motions more than 100 million years into the future.

Given their present trajectories across the globe, it appears that all the continents are headed for another collision. A quarter-billion years from now (more or less), most of Earth's land will again form one giant supercontinent—a land already named Novopangaea by some prescient geologists. However, the exact arrangement of that future is still a matter of debate.

Assembling Novopangaea is a tricky game. It's easy to take today's continental movements and predict ten or twenty million years down the road. The Atlantic will have widened by several hundred miles, while the Pacific will have shrunk by an equal amount. Australia will have moved north toward South Asia, and Antarctica will have shifted slightly away from the South Pole, also in the direction of South Asia. Africa is also on the move, inching northward to close off the Mediterranean Sea. In a few tens of millions of years, Africa will have collided with southern Europe, in the process closing up the Mediterranean and pushing up a Himalayan-size mountain range that will dwarf the Alps. So the map of the world twenty million years hence will appear familiar but skewed. Looking as far as one hundred million years into the future in this way is fairly safe, and most modelers arrive at similar geographics of a world where the Atlantic Ocean has overtaken the Pacific as the grandest body of water on Earth.

From that point on, models diverge. One school of thought, called extroversion, assumes that the Atlantic will continue opening and the Americas will eventually crunch into Asia, Australia, and Antarctica. In the latter stages of this supercontinent assembly, North America sweeps in from the east to close the Pacific and impact Japan, while South America wraps around clockwise from the southeast to snuggle against equatorial Antarctica. It's amazing how well all the pieces seem to fit together. Novopangaea is projected to be one immense landmass stretching east to west along the Equator.

The central assumption of this extroversion vision is that the great mantle convection cells that underlie plate motions will continue more or less as they do today. The alternative view, called introversion, takes the opposite tack by invoking past cycles of the opening and closing of the Atlantic Ocean. Reconstructions of the past billion years suggest that the Atlantic (or an equivalent ocean positioned between the Americas on the west and Europe plus Africa on the east) has opened and closed three times on a cycle of a few hundred million years—a result that suggests variable and episodic mantle convection. The rock record reveals that the movements of Laurentia and other continents about 600 million years ago formed the Atlantic's predecessor, called the Iapetus Ocean (named for the Greek Titan Iapetus, father of Atlas). The Iapetus Ocean closed at the assembly of Pangaea. When that supercontinent began to split apart 175 million years ago, the Atlantic Ocean was formed.

According to the introversion advocates (probably best not to call them introverts), the still-widening Atlantic will follow the same pattern. It will slow, stop, and reverse in about 100 million years. Then some 200 million years later, the Americas will once again collide with Europe and Africa. At the same time, Australia and Antarctica will have been sutured to Southeast Asia to complete the future supercontinent of "Amasia." This great landmass, shaped something like a sideways L, uses the same puzzle pieces as Novopangaea but this time with the Americas forming the western side.

For the time being, both supercontinent models, extroversion and introversion, appear to have merit and are still in play. And whatever the outcome of this friendly debate, everyone agrees that Earth's geography 250 million years from now, while strikingly different from today, will echo past times. The transient collection of continents at the Equator will reduce the impacts of ice ages and moderate changes in sea levels. Mountains will rise where continents collide, while patterns of weather and vegetation shift and atmospheric levels of carbon dioxide and oxygen fluctuate. Such changes will continue to be central to the story of Earth.

Impact: The Next Fifty Million Years

A recent survey of how people are most likely to die rated asteroid impacts pretty low—something like 1 in 100,000. That's statistically about the same probability as death by lightning or a tsunami. But there's an obvious flaw in this predictive comparison. Lightning kills one person at a time about sixty times per year. Asteroid impacts, by contrast, probably haven't killed anyone in thousands of years. But one really bad day, one little thwack could kill almost everyone all at once.

Chances are excellent that you don't have to worry, nor most likely will any of the next hundred generations. But we can be absolutely sure that another big impact of the dinosaur-killing variety is coming someday, somewhere. In the next fifty million years, Earth will suffer at least one big hit, maybe more. It's all a matter of time and probability. The most likely culprits are so-called Earth-crossing asteroids—objects with highly elliptical orbits that cross the plane of Earth's more circular path around the Sun. At least three hundred of these potential killers are known, and in the next few decades some of them will pass uncomfortably close. On February 22, 1995, a just-discovered asteroid with the benign name 1995 CR whizzed by within a few Earth-Moon distances. On September 29, 2004, asteroid Toutatis, an elongated 1.5-by-3 mile object, passed even closer. And in 2029 asteroid Apophis, a 900-foot diameter rock, is predicted to cross much closer still, well inside the Moon's orbit. That unsettling encounter will irrevocably alter the Apophis orbit and possibly bring it even closer in the future.

For every known Earth-crossing asteroid, there are probably a dozen or more yet to be spotted. And when one of these projectiles is finally observed, it will likely be much too close for us to do much about it. If we're the bull's-eye, we may only have a few days' warning to settle our affairs. Dry statistics tell the tale of probabilities. Earth is struck by a twenty-five-foot rock almost every year. Thanks to the braking effects of our atmosphere, most such missiles explode and fragment into little pieces before hitting the surface. But objects a hundred feet or more across, which arrive about once every thousand years, cause significant local damage: in June 1908, such an impactor leveled a swath of forest near the Tunguska River in Russia. Exceedingly dangerous half-mile stones impact on average about once every half-million years, while asteroids as large as three miles in diameter may arrive about once every ten million years.

The consequences of an impact will vary according to the size and location of the impact. A ten-mile boulder would devastate the globe just about anywhere it hits. (By contrast, the dinosaur-killing asteroid of 65 million years ago is estimated to have been about six miles across.) If a ten-mile object hits the oceans—a 70 percent chance, given the distribution of land and sea—then all but Earth's highest mountain peaks will be swept clean by immense globe-destroying waves. Nothing will survive up to a few thousand feet above sea level. Every coastal city will utterly disappear.

If such a ten-mile asteroid hits land, the immediate devastation may be more localized. Everything within a thousand miles would be obliterated, and massive fires would sweep across whatever continent is the unlucky target. For a short time, more distant lands might be spared the violence, but such an impact would vaporize immense quantities of rock and soil, sending Sun-obscuring clouds into the high atmosphere for a year or more. Photosynthesis would all but shut down. Plant life would be devastated and the food chain would collapse. A few humans might survive the horror, but civilization as we know it would be destroyed.

Smaller impactors would cause less death and destruction, but any asteroid over a few hundred feet, whether it smacks the land or the sea, would cause a natural disaster greater than anything we have known. What to do? Should we ignore the threat as too remote, too insignificant in a world that has so many more immediately pressing problems? What could we do to divert a big rock?

The late Carl Sagan, perhaps the most charismatic and influential spokesperson for science in the last half century, thought a lot about asteroids. In public and private, most famously in his epic TV series *Cosmos,* he advocated concerted international action. He set the stage by telling the vivid tale of the monks of Canterbury Cathedral, who in the summer of 1178 saw a violent explosion on the Moon—an asteroid impact so very close to us less than a thousand years ago. If such an impact occurred on Earth, countless millions would die. "The Earth is a very small stage in a vast cosmic arena," he said. "There is no hint that help will come from elsewhere."

The simplest, first step in avoiding such an event is to look as hard as we can for those elusive Earth-crossing destroyers—to know the enemy. We need dedicated telescopes, automated with digital processors, to locate the Earth-crossing projectiles, to plot their orbits and predict their future pathways. Such an endeavor is relatively cheap and already under way. More could be done, but at least the effort is being made.

And what if we found a large rock that is projected to smash into us a few years from now? For Sagan, along with others in both the scientific and the military communities, asteroid deflection is an obvious strategy. If initiated early enough, even a small nudge by a rocket engine or a few well-placed nuclear explosions could shift an asteroid's orbit sufficiently to change a collision course to a near-miss. Such an eventual necessity is reason enough for a robust program of space exploration, he argued. In a prescient 1993 essay, Sagan wrote, "Since hazards from asteroids and comets must apply to inhabited planets all over the Galaxy, if there are such, intelligent beings everywhere will have to unify their home worlds politically, leave their planets, and move to small nearby worlds around. Their eventual choice, as ours, is spaceflight or extinction."

Spaceflight or extinction. To survive in the long run, we must journey outward to colonize neighboring worlds. First will come bases on the Moon, though our luminous satellite will long remain a hostile place to live and work. Next is Mars, where more abundant resources—especially lots of frozen subsurface water, but also sunlight, minerals, and a tenuous atmosphere—are at hand. It won't be easy or cheap; nor is Mars destined to become a thriving colony anytime soon. But settling, and perhaps terra-forming, our promising neighbor may well be the next essential step in our species's evolution.

Two obvious obstacles will probably delay, if not prevent, the establishment of a Mars base. The first is money. The many tens of billions of dollars it will take to design and implement a Mars landing is outside the most optimistic NASA budget, even in the best of financial times. A cooperative global effort may be the only option, but such a massive international program has never been attempted.

Astronaut survival is an equally daunting challenge, for it's next to impossible to ensure a safe round-trip to Mars. Space is harsh, with myriad sand-size meteorite bullets to pierce the thin shell of even the most armored capsule and unpredictable solar bursts of lethal penetrating radiation. The Apollo astronauts, with their weeklong voyages to the Moon, were damned lucky that nothing bad happened. But a voyage to Mars would take many months; in the gamble inherent in any space mission, more time means more risk.

What's more, no rocket technology on the books would allow a spaceship to carry enough fuel to get to Mars and make it back. Some inventors talk of processing Martian water to synthesize enough fuel to refill the tanks, but that technology is only a dream and probably a long way off. Perhaps the more logical option—one that flies in the face of NASA mores but is increasingly promoted in passionate editorials—is a one-way trip. Were we to send an expedition with years of supplies instead of fuel, with sturdy shelter and a greenhouse, with seeds, with a lot of oxygen and water, and with tools to extract more life-giving resources from the red planet, then an expedition might just make it. It would be unbelievably dangerous, but so were many of the pioneering human voyages of discovery—the circumnavigation of Magellan in 1519–21, the westward explorations of Lewis and Clark in 1804–6, and the polar expeditions of Peary and Amundsen early in the twentieth century. Humans have not lost their lust to engage in such risky ventures. If NASA posted a sign-up sheet for the chance at a one-way trip to Mars, thousands of scientists would volunteer in a heartbeat.

Fifty million years from now, Earth will still be a vibrant living world, its blue oceans and green continents shifted but recognizable. The fate of our human species is much less certain. Perhaps we will be extinct. If so, fifty million years is more than sufficient to erase almost every trace of our brief dominion—every city, every highway, every monument would have weathered away millions of years earlier. Alien paleontologists would have to search long and hard to find the slightest near-surface trace of our vanished species.

But it is also possible that humans will survive and evolve, moving outward to colonize first our neighboring planets, then our neighboring stars. If so, if our descendants make it into space, then Earth will surely be treasured as never before—as a preserve, as a museum, as a shrine and place of pilgrimage. Perhaps only by leaving our world will humans ever fully appreciate the place of our species's birth.

Earth's Changing Map: The Next Million Years

In many respects, Earth a million years from now will not have changed very much. The continents will have shifted, to be sure, but probably no more than thirty or forty miles from their present relative locations. The Sun will still shine, rising every twenty-four hours, and the Moon will still orbit once every month or so.

But some things will have changed a lot. In many places around the globe, inexorable geological processes will have transformed the landscape. The most obvious changes will affect vulnerable coastlines. One of my favorite places, Calvert County, Maryland, with its miles of fast-eroding Miocene Epoch cliffs and their seemingly limitless store of fossils, will have disappeared completely. After all, the county is only five miles wide, and it's getting skinnier by almost a foot a year. At that rate, Calvert County won't last fifty thousand years, much less a million.

In other states, geological processes will add valuable new real estate. A new ocean-floor volcano off the southeastern coast of Hawaii's big island is already almost two miles high (though still submerged) and growing larger every year. A million years from now a new island, already named Loihi, will have risen from beneath the waves. Of course, older extinct volcanic islands to the northwest, including Maui, Oahu, and Kauai, will become correspondingly smaller as wind and waves erode them away.

Speaking of waves, scientists who examine the rock record for hints of what is to come conclude that Earth's geography will be altered most dramatically by the advancing and receding oceans. Changes in the rate of rift volcanism have a long-term effect, as greater or lesser volumes of lava solidify on the ocean floor. Sea level can drop significantly during lulls in ocean volcanism, when seafloor rocks cool and settle, as many suspect occurred during the dramatic fall of sea level just before the late Mesozoic extinction. The presence or absence of large inland seas like the Mediterranean and the assembly and breakup of continents cause major changes in the extent of shallow coastal waters, which play yet another important role in the shape the geosphere and biosphere will take in the next million years.

A million years corresponds to tens of thousands of human generations—hundreds of times all of recorded human history. If we survive, then Earth may be physically transformed by our evolving technological prowess in ways we cannot easily imagine. But if we die out, Earth will likely soldier on much as it is today. Life on land and in the sea will flourish; the coevolution of the geosphere and biosphere will quickly return to its preindustrial equilibrium.

Megavolcano: The Next Hundred Thousand Years

The sudden catastrophe of an asteroid impact may pale beside the time-released death of a megavolcano or flood basalt. Accompanying all of Earth's five greatest mass extinction intervals—including the one contemporaneous with a big rock falling from the sky—was globe-altering volcanism.

This is not to be confused with run-of-the-mill volcanic death and destruction in its various forms. These include dramatic lava flows like those so familiar to Hawaiian islanders living on the slopes of Kilauea—utterly destructive to any dwellings in their paths, but also localized, predictable, and easily avoided. Somewhat more deadly in this league of ordinary volcanism are the explosions and ash falls of pyroclastic volcanoes, which can release immense quantities of incandescent, steam-driven ash that race down a mountainside at more than a hundred miles per hour, incinerating and burying everything in their path. Cue the 1980 explosion of Mount St. Helens in Washington State and the June 1991 eruption of Mount Pinatubo in the Philippines, both of which would have killed thousands of people if prior warnings hadn't prompted mass evacuations. Still more ominous is a third type of volcanism as usual: the ejection of large quantities of fine-grained ash and toxic gases into the high atmosphere.

The Icelandic ash eruptions of Eyjafjallajökull in April 2010 and Grímsvötn in May 2011 were relatively puny, releasing much less than a cubic mile of debris. Nevertheless, they disrupted European air travel for several days and caused health concerns for many people in nearby regions. The eruption of Laki in June 1783—among the largest historic eruptions—released an estimated five cubic miles of basalt and associated ash and gas, sufficient to induce a long-lasting poisonous haze over Europe. A quarter of Iceland's population died, some rapidly from exposure to acidic volcanic gases and many more from starvation during the subsequent winter. The disaster also affected lands more than a thousand miles to the southeast, as tens of thousands Europeans, most in the British Isles, also died from Laki's prolonged effects. Even more people died following the August 1883 explosion of Krakatoa and the resultant tsunami that swept nearby coastlines of Java and Sumatra. And the colossal April 1815 eruption of Tambora, which produced an amazing twelve cubic miles of lava, was the deadliest of all. More than seventy thousand lives were lost, most as a consequence of agricultural failure and subsequent mass starvation. Tambora's injection of immense quantities of sun-blocking sulfur compounds into the upper atmosphere turned 1816 into the Northern Hemisphere's "year without a summer."

Such historic eruptions trouble the modern imagination, and for good reason. Sure, their death tolls pale by comparison with the hundreds of thousands of individuals killed by recent earthquakes in the Indian Ocean and Haiti. But there's an important, terrifying difference between earthquakes and volcanoes. The size of the largest possible earthquake is limited by the strength of rock. Hard rock can take only so much stress before it snaps; that extreme limit can produce an extremely destructive, but localized, earthquake—magnitude nine on the Richter scale.

Volcanoes, by contrast, have no apparent upper limit in size. In fact, the geological record holds unambiguous evidence of eruptions a hundred times greater than the largest volcanic events in recorded human history. Such megavolcanoes would have darkened the world's skies for years and altered the landscape over *millions* of square miles, not thousands. The most recent megavolcano explosion, Taupo on the North Island of New Zealand 26,500 years ago, may have produced more than 200 cubic miles of lava and ash. Toba in Sumatra, which erupted 74,000 years ago, released an estimated 672 cubic miles of ejecta. The consequences of another such catastrophe on modern society are hard to fathom.

And yet, even these megavolcanoes, though much greater than any cataclysm in recorded history, were dwarfed by the great flood basalts that contributed to mass extinctions. Unlike one-off explosions of megavolcanoes, flood basalts represent a sustained interval with thousands of years of intense volcanic activity. The greatest of these episodes, all of which coincide with global mass extinctions, produced hundreds of thousands to millions of cubic miles of lava. The biggest known event, now revealed by more than a half-million square miles of basalt flows, occurred in Siberia during Earth's greatest mass extinction, the great dying 251 million years ago. The demise of the dinosaurs 65 million years ago, so often ascribed to an asteroid impact, is also coincident with immense flood basalts in India—the Deccan Traps, almost 200,000 square miles in extent, representing more than 120,000 cubic miles of new rock.

These vast surface features could not have come from simple reprocessing of the crust and upper mantle. Current models of flood basalt formation envision a throwback to Earth's earliest age of vertical tectonics, with giant bubbles of magma slowly rising all the way from the superheated core-mantle boundary, cracking the crust, spewing out over the cold surface. Such events are now rare. One scenario posits a roughly thirty-million-year interval between flood basalts, in which case we're somewhat overdue for the next big one.

Our technological society will certainly receive fair warning of such an event. Seismologists will be able to track the hot, molten plume as it rises. We may have hundreds of years to prepare for the calamity. But should humanity ever enter another era of megavolcanism, there would be nothing we could do to stop Earth's most violent paroxysms.

Ice Factor: The Next Fifty Thousand Years

For the foreseeable future, the biggest determining factor in Earth's continental contours is ice. On short timescales of a few hundred or thousand years, the depth of the oceans is most closely tied to the total volume of Earth's frozen water, including the ice caps, glaciers, and continental ice sheets. It's a simple equation: the greater the volume of water tied up in ice on land, the lower the level of the sea.

The past is key to predicting the future, but how can we possibly know the depth of historic oceans? Satellite observations of ocean levels, though incredibly accurate, are restricted to about the past two decades. Tide gauge measurements, though less accurate and subject to local idiosyncrasies, go back perhaps a century and a half. Coastal geologists can resort to mapping ancient shoreline markers—raised beach terraces, for example, which can be found in near-shore accumulations of sediments dating back tens of thousands of years, though such elevated features can reliably reveal only periods of higher water levels. The positions of fossil corals, which must have grown in the ocean's shallow sunshine zone, can push the record even further back, but such rock formations commonly experience episodes of uplift, subsidence, or tilting that confuse the record.

Many scientists now focus on a less obvious indicator of sea level— the variable ratio of the isotopes of oxygen in tiny seashells. Such ratios tell us much, much more than a cosmic body's distance from the Sun, as discussed in chapter 2. Because of their temperature-sensitive nature, oxygen isotopes are also the key to deciphering the historic volume of Earth's ice and thus ancient sea level changes.

Even so, the connection between ice volume and oxygen isotopes is tricky. By far the most abundant oxygen isotope, accounting for about 99.8 percent of what we breathe, is the lighter oxygen-16 (with eight protons and eight neutrons). About one in five hundred oxygen atoms is the heavier oxygen-18 (with eight protons and ten neutrons). That means about one in every five hundred water molecules in the ocean is heavier than the average. As the Sun heats equatorial ocean water, water with the lighter oxygen-16 isotope evaporates a bit faster than that with oxygen-18, ensuring that water in low-latitude clouds is a bit lighter on average than the oceans from which it came. As the clouds rise to cooler zones, water with the heavier oxygen-18 isotope condenses into raindrops a bit faster than that with oxygen-16, ensuring that the cloud's oxygen becomes even lighter than it was before. By the time the clouds move toward the poles, as clouds inevitably do, the oxygen in their water molecules has become much lighter than in those of ocean water. When these polar clouds release their precipitation onto ice caps and glaciers, more of the lighter isotope is locked in the ice, leaving the oceans heavier.

During times of maximum global cooling, when more than 5 percent of Earth's water can be frozen solid, the oceans become significantly enriched in oxygen-18. During times of global warming and glacial retreat, oxygen-18 levels in the oceans decline. So it is that careful layer-by-layer measurements of oxygen isotope ratios in coastal sediments can reveal how Earth's surface ice (or lack thereof) has varied through time.

Such exacting work is the domain of geologist Ken Miller and his colleagues at Rutgers University, who have spent decades scrutinizing the thick accumulations of marine sediments that blanket coastal New Jersey. These sediments, with a record extending back one hundred million years, contain a wealth of microscopic fossil shells called foraminifera, or forams for short. Each tiny foram preserves the oxygen isotope content of the ocean at the time when it grew. Layer-by-layer measurements of oxygen isotopes in New Jersey sediments thus provide a simple and accurate estimate of the ice volume through time.

In the recent geological past, the ice cover appears to have ebbed and flowed constantly, with correspondingly large changes in sea level on a timescale of a few thousand years. At the height of recent ice ages, more than 5 percent of Earth's water has been trapped in ice, lowering sea levels by perhaps three hundred feet from their present mark. About twenty thousand years ago, such a period of low ocean level is thought to have created a land bridge between Asia and North

America across what is now the Bering Strait—the original passage-way for humans and other mammals to the New World. During the same icy interval, there was no English Channel, as a dry valley connected the British Isles to France. By contrast, at times of maximum warming, when glaciers largely disappear and ice caps recede, sea levels have repeatedly risen by as much as three hundred feet higher than today, submerging hundreds of thousands of square miles of coastal areas around the globe.

Miller and his colleagues identified more than a hundred cycles of glacial advance and retreat in the past nine million years, with at least a dozen such events in just the past million years alone—changes that point to wild swings of as much as six hundred feet in ocean level. While details may vary from cycle to cycle, these events are clearly periodic and are related to so-called Milankovitch cycles, named for the Serbian astrophysicist Milutin Milankovitch, who discovered them about a century ago. He realized that well-known variations in Earth's orbit around the Sun, including our planet's tilt, its elliptical orbit, and a slight wobble in its rotation axis, impose periods of climate change with intervals of roughly twenty, forty-one, and one hundred thousand years. These variations all affect the amount of sunlight hitting Earth and thus exert a profound effect on global climate.

So what of the next fifty thousand years? We can be confident that sea levels will continue to vary dramatically, with many more rises and falls. At times, very possibly over the next twenty thousand years, ice caps will grow, glaciers will advance, and sea level will decrease by two hundred feet or more—a level that has occurred at least eight times over the past million years. Such a change will have powerful effects on the world's coastlines. The East Coast of the United States will shift many miles to the east, as the shallow continental slope is exposed. All of the great East Coast harbors, from Boston to Miami, will become high and dry inland cities. A new ice and land bridge will connect Alaska to Russia, and the British Isles may once again become a part of mainland Europe. Meanwhile the world's most productive fishing grounds along the continental shelves will become dry land.

In the case of sea level, what goes down must also go up. It's very possible, some would say very likely, that over the next thousand years sea level will increase by a hundred feet or more. Such a rise in the oceans, a rather modest one by geological standards, would render the map of the United States all but unrecognizable. A hundred-foot rise in sea level would inundate much of the coastal plain of the East Coast, shifting shorelines up to a hundred miles westward. All the major coastal cities—Boston, New York, Philadelphia, Wilmington, Baltimore, Washington, Charleston, Savannah, Jacksonville, Miami, and more—will be submerged. Los Angeles, San Francisco, San Diego, and Seattle will also disappear beneath the waves. Almost all of Florida will be gone, its distinctive peninsula drowned in a shallow sea. Most of Delaware and Louisiana will be under water, too. In other parts of the world, consequences of a hundred-foot rise in sea level will be even more devastating. Entire countries—the Netherlands, Bangladesh, the Maldives—will be no more.

The geological record is unambiguous: these changes will happen again. And if Earth is warming rapidly, as most experts suspect, then waters will rise soon, perhaps as rapidly as a foot per decade. Mere thermal expansion of ocean water during extended periods of global warming can increase average sea levels by up to ten feet. Such changes will challenge human society, for sure, but will have little effect on Earth.

After all, it wouldn't be the end of the world. Just our world.

Warming: The Next Hundred Years

Most of us don't care so much about a few billion years in the future, or a few million years, or even a thousand years. Most of us focus on short-term concerns: How will I pay for my kid's college in ten years? Will I get that promotion next year? Will the stock market go up next week? What's for dinner?

In that context, we have little to worry about. Barring an unforeseen cataclysm, Earth next year, next decade, will seem pretty much as it does today. Any differences from one year to the next will probably be too small to notice, even if we experience an uncharacteristically hot summer, endure a crop-withering drought, or suffer an unusually violent storm.

What is absolutely certain is that Earth will continue to change. Present indicators point to a coming episode of global warming and melting glaciers, most probably influenced and accelerated by human activities. Over the next hundred years, the consequences of this warming will affect many people in many different ways.

In the summer of 2007, I participated in a Kavli Futures Symposium in remote Ilulissat, a fishing village on the west coast of Greenland, a short distance south of the Arctic Circle. It was a fortunate choice for discussing the future, as changes were occurring immediately outside our conference center at the cozy Arctic Hotel. For a thousand years, the harbor, located near the calving front of the mighty Ilulissat Glacier, had served as a rich fishing ground. For a thousand years, the fishermen resorted to ice fishing in the winter as the harbor froze solid every year. Until the new millennium, that is. In 2000, for the first time (at least in a thousand years of oral history), the harbor was open and unfrozen. The mighty glacier, a United Nations heritage site, has been receding at an astonishing rate—almost six miles in three years after many decades of stability. Another change: for a thousand years Ilulissat and nearby native villages had been free of insect pests, but in 2007 and all subsequent years, an infestation of mosquitoes and black flies arrived in August. These are anecdotes, to be sure, but so too are they unambiguous harbingers of significant, inexorable change.

Across the globe, similar changes are occurring. Watermen on the Chesapeake Bay report consistently higher high tides than a few decades ago. Year by year the northern Sahara Desert is pushing ever farther north, turning once fertile Moroccan farmland into dust. Antarctic ice shelves are melting and breaking off at increased rates. Average global air and water temperatures are on the rise. It's all part of a consistent pattern of warming—a pattern that Earth has experienced countless times in the past and will experience countless times in the future.

Warming can have other, sometimes paradoxical effects. The Gulf Stream, the great ocean current that carries warm water from the Equator to the North Atlantic, is powered by the strong temperature differences between the Equator and higher latitudes. If global warming reduces that temperature contrast, as some climate models suggest, then the Gulf Stream may weaken or even stop. Ironically, an immediate consequence would be to make the British Isles and northern Europe, where climate is moderated by the Gulf Stream, much colder than it is today. Other ocean currents—for example, those from the Indian Ocean to the South Atlantic past the Horn of Africa—would be similarly affected and might cause a similar shift in the mild South African climate or a change in the monsoon rains that keep parts of Asia wet and fertile.

As ice melts, the seas rise. Some sober projections suggest increases of as much as two or three feet in the next century, though much faster increases of several inches per decade may have occurred from time to time according to the recent rock record. Such a sea change will affect many coastal residents around the world and may cause headaches for civil engineers and beachfront property owners from Maine to Florida, but a few feet is probably manageable in most populated coastal areas. For a while, for a generation or two, most residents really won't have much to worry about when it comes to encroaching seawater.

Some animal and plant species may not fare so well. The loss of northern polar ice will reduce the familiar habitat of polar bears, adding challenges to a population that appears to be shrinking. A rapid shift of climate zones toward the poles may also stress many other threatened species, notably birds, which are particularly susceptible to alterations in their migratory nesting and feeding areas. One recent report estimated that an average global temperature rise of just a couple of degrees, well within predictions of some climate models for the next century, could trigger extinction rates among birds approaching 40 percent in Europe and exceeding 70 percent in the lush rain forests of northeastern Australia. Another sobering international report found that nearly one in three of the approximately six thousand known species of frogs, toads, and salamanders is similarly endangered, mainly by the rapid warmth-driven expansion of a deadly amphibian fungal disease. Whatever else transpires over the coming century, it does appear that we are entering a time of accelerated extinction.

Certain transformative events of the next century—some guaranteed, others highly possible—will be instantaneous: the disruption of a great earthquake, the eruption of a megavolcano, or the impact of a mile-wide asteroid. Human societies tend to be ill prepared for the once-in-a-century storm or earthquake, much less for the truly catastrophic once-in-a-thousand-years disaster. As we read Earth's story, we see that these shocking events are the norm, inevitable, a part of the continuum of our planet's history. Nevertheless, we build our cities on the flanks of active volcanoes and on some of Earth's most active fault zones, hoping that in our time we will dodge the tectonic bullet (if not the cosmic missile).

In between the very slow and very fast are fluctuating geological processes that normally take hundreds or thousands of years—shifts in climates, sea level, and ecosystems that are usually noticeable only over the span of several generations. The *rates* of such changes, not the changes themselves, should be our biggest concern. For climate, sea level, and ecosystems can reach tipping points. Pushed too far, positive feedback loops can kick in. What normally takes a thousand years could transpire in a decade or two.

Complacency is easy, especially when bolstered by flawed readings of the rocks. For a while, until 2010, concerns about modern times were somewhat assuaged by ongoing studies of a parallel scenario 56 million years ago—one of the mass extinctions that dramatically affected the early evolution and spread of mammals. This harsh event, called the Paleocene-Eocene Thermal Maximum, or PETM for short, saw the relatively sudden disappearance of thousands of species. The PETM is important for our time because it is the most rapid well-documented temperature shift in Earth history. A relatively fast volcano-induced increase in concentrations of atmospheric carbon dioxide and methane, the twin heat-trapping gases of the greenhouse effect, caused more than a thousand years of positive feedbacks and a corresponding episode of modest global warming. Some researchers saw the PETM as a close parallel to today's events, bad, to be sure—with an almost 10-degree rise in global temperatures, a rapid rise in sea level, acidification of the oceans, and significant poleward shifts in ecosystems—but not so catastrophic as to threaten the survival of most animals and plants.

Shocking recent discoveries by Penn State geologist Lee Kump and his colleagues may have destroyed any lingering cause for optimism. In 2008 Kump's team was given access to a drill core from Norway that preserved the entire interval of the PETM—sedimentary rocks that, layer by layer, documented in exquisite detail the rates of change of atmospheric carbon dioxide and climate. The bad news is that the PETM—what has for more than a decade been thought to be the most rapid climate disruption in Earth's history—was triggered by atmospheric changes less than a tenth the intensity of what is happening today. Global changes in atmospheric composition and average temperature that took more than a thousand years during the PETM extinction scenario have been surpassed in just the last hundred years, as humans have burned immense quantities of carbon-rich fuels.

There is no known precedent for such rapid change, and no one knows how Earth will respond. At an August 2011 meeting of three thousand geochemists in Prague, the mood among climate specialists familiar with the new PETM data was sober. Though public predictions by these cautious experts have remained measured, the comments I heard over a beer were pessimistic, frightening. If greenhouse gas concentrations rise too rapidly, no known mechanisms can absorb the excess. Will warming trigger a massive methane release, with all the positive feedbacks that that scenario might entail? Will sea level quickly rise hundreds of feet, as it has so many times in the past? We are venturing into terra incognita—conducting an ill-conceived global-scale experiment on Earth perhaps unlike any that has come before.

What the testimony of the rocks does reveal is that, as resilient as life itself is and always will be, the biosphere experiences great stress at tipping points, during times of sudden climate shifts. Biological productivity, including agricultural productivity, will most certainly fall precipitously for a time. Under such dynamic conditions, large animals like ourselves will pay the dearest price. The coevolution of rocks and life will continue unabated, to be sure, but humanity's role in that multibillion-year saga remains unknowable.

Have we already reached such a tipping point? Probably not in this decade, maybe not in our lifetimes. But that's the thing about tipping points—you can never be sure you're at one until it's happened. The housing bubble bursts. The populace of Egypt revolts. The stock market crashes. We realize what's happening only in retrospect, when it's too late to restore the status quo. Not that there has ever been any such thing in the story of Earth.

Epilogue

∾ Climates change, sea levels change, the rains and winds change, the distributions of life across the surface and within the seas change. Rocks and life continue to coevolve, as they have for billions of years. Humans can no more stop global change than we can alter Earth's trajectory through the cosmos.

Nor can we destroy life on Earth, nor even stop its inexorable evolution. Life has ensconced itself in every niche of the globe. Life abounds in frozen Arctic ice, in boiling acid pools, in rock-encased pores miles underground, and on wind-borne dust specks miles above the ground. Whatever stupidity we might inflict upon ourselves—whether we cause global temperatures to rise a dozen degrees, or poison our air and water, or decimate fish stocks in the sea, or even unleash our collective nuclear arsenals in a global holocaust—life will go on. Humans may disappear forever, but microscopic life will scarcely miss a beat. For billions of years to come, Earth will continue to whirl daily on its axis in its annual odyssey about the Sun. For billions of years, ours will still be a living planet of blue oceans, green lands, and swirling white clouds. From space, Earth will be no less beautiful than it is today, humans or no.

Make no mistake. There cannot be the slightest doubt that human activities of the past century have initiated dramatic changes in atmospheric composition, and changes in climate must follow as surely as the laws of physics. The concentrations of carbon dioxide and methane, both efficient greenhouse gases, have climbed at rates unmatched in hundreds of millions of years. Such changes are amplified by our rapid denuding of tropical rain forests, our efficient consumption of sea life, and our incessant destruction of habitats across the globe. Thanks to our actions, Earth *will* get hotter, ice *will* melt, oceans *will* rise. But that's nothing new for Earth. Why, then, should we care if human actions speed up the process of change?

For one, imagine the suffering unleashed on a world where sea life undergoes mass death or agricultural output is suddenly halved. What of the million square miles of the most productive farmland that will be flooded, the seaports drowned, the livelihoods lost? Imagine the suffering of a billion displaced and homeless humans.

If we care to act, it is surely not in order to "save the planet." Earth, after surviving more than 4.5 billion years of constant, extravagant change, doesn't need saving. Perhaps some ethicists will focus their efforts instead on saving the whales or polar bears, for their loss would be permanent and undeniably sad. But even the extinction of these great beasts, or of elephants or pandas or rhinos or a million other species both charismatic and mundane, is but a temporary loss to Earth. New great and wondrous beasts will inevitably evolve, in a geological moment, to fill those vacant niches—perhaps in no more than a million years. Large mammals like ourselves may suffer mass extinction, but other vertebrates, maybe the birds, will take our place. Maybe penguins, which have recently been shown to evolve particularly rapidly, will morph and radiate to fill the niches: whalelike penguins, tigerlike penguins, and horselike penguins. Maybe the penguins will develop big brains and grasping fingers. Whatever we do, Earth will continue to be a variegated living world.

No, if we choose to worry, it should be first and foremost for our human family, for it is we who are most at risk. Earth is a great winnower of waste and error. Life in its grandeur will go on, but human society, at least in its present profligate mode, may not make the cut. We humans have the sobering potential, through either our thoughtless actions or our equally thoughtless inaction, to heap untold suffering and destruction upon our own species. As we continue to alter our home world—our "pale blue dot," to quote Carl Sagan—at a faster and faster rate, the time remaining for effective action slips away.

Earth is not silent on this point; its story is there to read in the rich record of the rocks. For thousands of years, we have been wise enough to seek out the story of Earth in an effort to know our home. Let us hope we learn its lessons in time.

Acknowledgments

ᔕ Dozens of friends and colleagues have contributed to the concept and development of this volume. I am especially indebted to four scientists who embraced the concept of mineral evolution in 2008, in its early stages. Mineralogist Robert Downs, a longtime friend and collaborator, provided his considerable expertise on the nature and distribution of minerals. Petrologist John Ferry of Johns Hopkins University, whom I've known since our days in graduate school, contributed a sophisticated theoretical framework for the new approach to mineralogy. Geobiologist Dominic Papineau, a former Geophysical Laboratory postdoctoral fellow and now on the faculty of Boston College, was one of the earliest contributors to and most perceptive and constructive critics of the mineral evolution idea, in spite of the objections of his other Carnegie mentors. Geochemist Dimitri Sverjensky of Johns Hopkins University, my closest professional colleague for the past several years, brought a wealth of ideas and insights to the conceptual development of mineral evolution. These four friends were the earliest champions of the mineral evolution idea, and all have been articulate and effective collaborators. This book would not have been possible without their help.

We gained invaluable insights from Precambrian geologist Wouter Bleaker of the Geological Survey of Canada, meteorite expert Timothy McCoy of the Smithsonian Institution, and biomineralogy authority

Hexiong Yang of the University of Arizona. They joined in the initial publication of these ideas. Subsequent collaborations with David Azzolini, Andrey Bekker, David Bish, Rodney Ewing, James Farquhar, Joshua Golden, Andrew Knoll, Melissa McMillan, Jolyon Ralph, and John Valley have amplified the concept and led us into exciting new directions. I am especially indebted to Edward Grew, whose studies of the evolution of the minerals of the rare elements beryllium and boron have taken the field to a new quantitative level.

This book could not have been undertaken were it not for my many colleagues in the origins-of-life field. Special thanks go to Henderson James Cleaves, George Cody, David Deamer, Charlene Estrada, Caroline Jonsson, Christopher Jonsson, Namhey Lee, Kataryna Klochko, Shohei Ono, and Adrian Villegas-Jimenez. I have also benefited immeasurably from collaborations with Harvard paleontologist Andrew Knoll and several of his associates, notably Charles Kevin Boyce and Nora Noffke, as well as Neil Gupta.

I have received unflagging support from my colleagues Connie Bertka, Andrea Magnum, and Lauren Cryan at the Deep Carbon Observatory, as well as Jesse Ausubel of the Alfred P. Sloan Foundation, which has provided generous support to launch this global effort. They have borne much of the brunt of my distractions while working on this book. My colleagues at George Mason University, especially Richard Diecchio, Harold Morowitz, and James Trefil, have engaged in numerous stimulating discussions throughout the development of the mineral evolution concept. I am also grateful to Russell Hemley, director of the Geophysical Laboratory, who has offered unqualified support and encouragement for this project.

Many scientists offered invaluable advice and information during the research for this book. I thank Robert Blankenship, Alan Boss, Jochen Brocks, Donald Canfield, Linda Elkins-Tanton, Erik Hauri, Linda Kah, Lynn Margulis, Ken Miller, Larry Nittler, Peter Olson, John Rogers,

Hendrick Schatz, Scott Shepard, Steve Shirey, Roger Summons, and Martin van Kranendonk.

This softcover edition has also benefitted from the thoughtful comments and corrections of Ariel Anbar, Chris McKay, and Bruce Sherwood. I am grateful to the Viking editorial and production team for their enthusiasm and professionalism in the development of this book. Alessandra Lusardi first championed the book and provided critical advice in its developmental phase. Liz Van Hoose provided invaluable editorial guidance and ushered the manuscript to its final state with creativity, efficiency, and good humor. I'd also like to thank Bruce Giffords and Janet Biehl.

The original idea for this book was developed in collaboration with Eric Lupfer of William Morris Endeavor, who has provided thoughtful analysis, timely advice, and constant support at every stage of this project. I am much in his debt.

Margaret Hazen has helped me throughout the development of the mineral evolution idea, from long before its first articulation on December 6, 2006, to its presentation in this volume. Her keen eye and infectious enthusiasm in the field, her informed advice and incisive critiques on all my manuscripts, and her spontaneous joy and gentle sympathy in response to the successes and failures of an intense research career have sustained this effort.

Index